HURRICANE ANDREW

Hurricane Andrew has proved to be the most costly natural disaster in US history. This book documents how Miami prepared for, coped with and responded to the hurricane, which slammed into one of the largest and most ethnically diverse metropolitan areas of the United States – Miami. With sustained winds of 145 mph, the infrastructure in the southern metropolitan area was laid to waste – nearly all public buildings were severely damaged or destroyed. Approximately 49,000 private homes were rendered uninhabitable, leaving more than 180,000 people homeless. Total losses were in excess of $28 billion.

This book explores how social, economic and political factors set the stage for Hurricane Andrew by influencing who was prepared, who was hit the hardest, and who was most likely to recover. Disasters are often seen as natural physical phenomena that impact our communities in impartial ways. As a result, the damage they inflict and the difficulties experienced in recovering are simply seen as a function of the strength of the agent itself and where it happens to hit the hardest. But disasters are inherently social events. The nature of our communities – how they are organized, how they exploit and use the natural environment and how scarce resources such as housing are distributed – is a critical factor for understanding disaster impact and recovery.

Employing data they collected over three years using qualitative and quantitative techniques, the authors of *Hurricane Andrew* analyze the consequences of conflict and competition, especially those associated with race, ethnicity and gender, on preparation, response and recovery.

Walter Gillis Peacock is Director of Research at the International Hurricane Center, Florida International University, and **Betty Hearn Morrow** and **Hugh Gladwin** are Associate Professors of Sociology and Anthropology at Florida International University.

HURRICANE ANDREW

Ethnicity, gender and the sociology
of disasters

Edited by
Walter Gillis Peacock,
Betty Hearn Morrow and
Hugh Gladwin

Routledge
Taylor & Francis Group

LONDON AND NEW YORK

First published 1997
by Routledge
2 Park Square, Milton Park, Abingdon, Oxfordshire OX14 4RN

Simultaneously published in the USA and Canada
by Routledge
711 Third Avenue, New York, NY 10017

First issued in paperback 2014

Routledge is an imprint of the Taylor & Francis Group, an informa business

Typeset in Garamond by Keystroke, Jacaranda Lodge,
Wolverhampton

British Library Cataloguing in Publication Data
A catalogue record for this book is available from the British Library

Library of Congress Cataloguing in Publication Data
Hurricane Andrew : ethnicity, gender and the sociology of disasters / edited by
Walter Gillis Peacock, Betty Hearn Morrow and Hugh Gladwin.
p.   cm.
Includes bibliographical references and index.
1. Hurricane Andrew, 1992. 2. Disaster relief–Florida–Dade
County. 3. Emergency management–Florida–Dade County.
4. Hurricane protection. I. Morrow, Betty Hearn, 1940– .
II. Gladwin, Hugh. III. Title.
HV636 1992. F672 1997
363.34′9228′0975938–dc21   97–7146

ISBN 13: 978-1-138-86711-6 (pbk)
ISBN 13: 978-0-415-16811-3 (hbk)

Dedicated to
Frederick L. Bates,
pioneer in the application of
socio-political ecology
to the study of
human systems and disasters

CONTENTS

CONTENTS

PLATES

PLATES

FIGURES

TABLES

TABLES

NOTES ON CONTRIBUTORS

Harvey Averch is professor of public administration at Florida International University and holds a Ph.D. in economics from the University of North Carolina. His most recent book is *Private Markets and Public Intervention* (Pittsburgh 1990).

Nicole Dash is a research associate with the International Hurricane Center while completing her doctorate in Comparative Sociology at Florida International University. Her interests include the appropriate application of technology to disaster management and response, and she has served as a GIS specialist for FEMA and as a research assistant at the Disaster Research Center at the University of Delaware.

Milan J. Dluhy is professor of public administration and Director of the Institute of Government at Florida International University. He received his Ph.D. from the University of Michigan. Since 1988 he has published three books on government effectiveness in service provision.

Elaine Enarson is a sociologist affiliated with the University of British Columbia Centre for Disaster Preparedness and Resources and the UBC Centre for Research in Women's Studies and Gender Relations. Her current research interests include violence against women in disaster contexts and gender relations in emergency management. She is co-editor, with Betty Hearn Morrow, of *The Gendered Terrain of Disasters: Through Women's Eyes* (Greenwood, forthcoming).

Chris Girard is director of the comparative sociology graduate program at Florida International University. His research interests include race and ethnicity, global patterns of suicide, medical sociology and segregation. His work has appeared in the *American Sociological Review* and the *American Journal of Sociology*, among others.

Guillermo J. Grenier serves as director of the Center for Labor Research and Studies at Florida International University. He is co-editor of *Miami Now! Immigration, Ethnicity, and Social Change* (University Press of Florida 1992) and a co-author of *Newcomers in the Workplace: Immigrants and the Restructuring of* *the US Economy* (Temple 1993) and *This Land is Our Land: Immigrants and Established Residents in Miami* (Temple 1997).

Hugh Gladwin is director of the Institute for Public Opinion Research and associate professor of anthropology at Florida International University. As a research associate of the International Hurricane Center he is involved in vulnerability mapping, and hurricane evacuation and impact modeling.

Betty Hearn Morrow is associate professor of sociology and a research associate of the International Hurricane Center at Florida International University. Her current research interests include family and gender issues associated with disasters. She is co-editor, with Elaine Enarson, of *The Gendered Terrain of Disasters: Through Women's Eyes* (Greenwood, forthcoming).

Walter Gillis Peacock is associate professor of sociology and director of research at the International Hurricane Center. He is co-author, with Frederick L. Bates, of *Living Conditions, Disasters and Development* (University of Georgia Press 1993) and his work has appeared in the *American Sociological Review* and the *International Journal of Mass Emergencies and Disasters*, among others.

A. Kathleen Ragsdale is currently pursuing a doctorate in anthropology at the University of Florida. At the time of this work she was a master's student and research assistant at Florida International University. Her principal interests include visual anthropology and gender.

Kevin A. Yelvington is associate professor of anthropology at the University of South Florida. His research interests include ethnic, class and gender relations and applied anthropology. His research focuses on Latin America and the Caribbean and he is the author of *Producing Power: Ethnicity, Gender, and Class in a Caribbean Workplace* (Temple 1995).

ACKNOWLEDGMENTS

Our research efforts began with support from Florida International University's President Modesto (Mitch) Maidique, the Division of Sponsored Research and Training (DSRT), and the College of Arts and Sciences. Additional research support in the form of grant and contract monies were obtained from the National Science Foundation (SES-9224537), the University of Florida's Bureau of Economic and Business Research, the American Red Cross, the Planning Department of the City of Homestead, the Florida International University Foundation and the John S. and James L. Knight Foundation. Of course, any opinions, findings, conclusions or recommendations expressed in this book are those of the editors and authors and do not necessarily reflect the views of these agencies or organizations.

As editors we wish to acknowledge and express our deepest appreciation to the faculty and graduate students who were members of the FIU Disaster Research Team. They include: Manny Alba (now a doctoral student at the University of Oklahoma), Linda Beer (now a doctoral student at Emory University), Nicole Dash (Research Associate, FIU International Hurricane Center), Elaine Enarson (now a sociologist at the Centre for Research in Women's Studies and Gender Relations at the University of British Columbia), Chris Girard (Associate Professor of Sociology, FIU), Guillermo Grenier (Associate Professor of Sociology and Director of the Center for Labor Research and Studies, FIU), Donna Kerner, Barry Levine (Professor of Sociology, FIU), Kathleen Ragsdale (now a doctoral student at the University of Florida), and Kevin Yelvington (now an Assistant Professor of Anthropology, University of South Florida).

We are most grateful for the time and information shared with us by thousands of victims and responders. The insights and understandings we gained would not have been possible without their willingness to participate, often under very difficult circumstances.

We wish to acknowledge a number of individuals who assisted in obtaining funding, gaining access to information or otherwise provided their unique insights and support. They include: Juanita and Steve Mainster (Centro Campesino), Beth von Werne (Catholic Services and later Lutheran Ministries),

ACKNOWLEDGMENTS

Douglas McLaughen and Sandy Kief (FIU Institute for Public Opinion Research), Kate Hale (Kate Hale and Associates, formerly Dade County Emergency Manager), Fred Murphy (Metro Dade Emergency Management), Kay Hunley (American Red Cross), Thomas Wilson (FIU/FAU Joint Center for Environmental and Urban Problems), Tom Breslin and Kathy Thurman (Division of Sponsored Research and Training, FIU), Peggy Nolan (Photography, FIU), David Lawson (FEMA Region IV GIS), Anthony Oliver-Smith (Anthropology, University of Florida), James Ito-Adler (Sociology and Anthropology, FIU), Sara Tardanico, Rick Tardanico (FIU Sociology and Anthropology, FIU), Chuck Blowers and Oliver Kerr (Metro Dade Planning Department). We are very grateful to Michelle Lamarre for indispensable help at keeping track of the accounts.

Frederick L. Bates, of the University of Georgia, must be accorded a special acknowledgment. While Fred would probably argue – with or without a bottle of Scotch – that we got many things wrong, his ideas and work were the foundation upon which much of this is based. In recognition of his contributions to the field, and to our professional development, it is dedicated to him.

Our appreciation is extended to Dave Lawrence, Publisher and Chairman, and the staff of the *Miami Herald*. They allowed full access to their fine collection of photos documenting the aftermath of Hurricane Andrew, many of which are included in this book. Equally important was the astute and in-depth reporting of the complexities associated with South Florida's experiencing of this event which provided us with a relevant stream of fresh information and data. The awards the *Miami Herald*, its reporters and photographers received following Hurricane Andrew do not begin to adequately recognize their extraordinary work in bringing to light not just the physical event, but also the full scope of its social and political consequences. The *Herald* has played and must continue to play a pivotal role in educating the public if we are to be better prepared the next time.

Finally, we want to offer a special thanks and acknowledgment to our families, who have had to endure our obsession with this research and this book. Eve, Bert, and Mary, you may hope that the *ménage à trois* is over, but the next project looms on the horizon.

ABBREVIATIONS

ARC	American Red Cross
BEBR	Bureau of Economic and Business Research
CDBG	Community Development Block Grant
DAC	Disaster Assistance Center
DIRC	Disaster Information and Response Center
EOC	Emergency Operations Center
FEMA	Federal Emergency Management Agency
FIRE	Finance, Insurance and Real-Estate
FIU	Florida International University
GIS	geographical information systems
HRS	Health and Rehabilitative Services
IA	Individual Assistance
ICARE	Interfaith Coalition for Andrew Recovery Effort
IFG	Individual Family Grant
NHC	National Hurricane Center
NOAA	National Oceanic and Atmosphere Administration
NSDPC	New South Dade Planning Charrette
PTSD	post traumatic stress disorder
SBA	Small Business Administration
SOPs	standard operating procedures
VISTA	Volunteers in Service to America
VOAD	Voluntary Organizations Active in Disasters
VOLAG	Voluntary Agencies
WWR	We Will Rebuild

1

DISASTER IN THE FIRST PERSON

Betty Hearn Morrow

It was a dark and stormy night . . . [1]

Considered in literary circles to be the best example of a bad opening line, this hokum phrase somehow seems appropriate. In preparing this account of the consequences of Hurricane Andrew on the people of South Florida, we have encountered, as well as personally experienced, a full range of human emotions – beginning with the fear of that night and continuing through often unpredictable vacillations between depression and optimism, joy and despair. Throughout the long process of recovery, however, has run a thread of courage, often expressed as humor. We first noticed it in the hundreds of messages, such as "For sale – half off," "Life begins at 165 mph," "Looters will be eaten," "Is this the best you've got, Andy?" and "Customized by Andrew," defiantly painted on the walls of destroyed homes. We saw it again in the whimsical way some homeless victims decorated the fronts of their tents as though they were palace entrances. In that same vein, when learning about our book project, several Andrew survivors wryly suggested this opening line.

There is little need for an explanation as to how we became interested in this project. We lived it. Only two of us had previously worked in the sociology of disasters field and only one, Walt Peacock, had extensive experience and training in the area. I had completed a small project on the social effects of Hurricane Hugo on St. Croix several years ago. Walt's prior research had included the long-term consequences of disasters and aid delivery programs in developing countries, particularly Guatemala. Ironically, having just moved to Miami to join the faculty at Florida International University when Hurricane Andrew hit, he was often asked why he had to bring his work with him. From its impact, there was no doubt that we would analyze the social impact of this hurricane. We had yet to realize, however, that we were in the middle of what would turn out to be the most destructive natural disaster the United States had experienced.

Our first trip deep into the most southern part of Dade County was in a police car, since normal transportation was quite impossible. The devastation was beyond our worse expectations. Soon we were observing and interviewing

1

in the tent cities established by the military to house thousands of homeless victims. For our team this was the beginning of three years of data collection and immersion into a difficult and seemingly endless project. We hope the results do justice to the story of the thousands of people – victims and responders – who have enlightened us along the way. It is also important that the projects chronicled in this book make a contribution to the growing body of knowledge on the social impact of disasters and, in some small way, enlighten efforts to mitigate the effects of future events. Thus, we end most chapters with a discussion of the policy implications of our work.

THE EVENT

When first named, Andrew did not make much of an impression. On Friday 21 August 1992 (three days before it would make landfall) the National Hurricane Center advisories referred to it as barely a tropical storm moving slowly across the Atlantic. It was expected to blow apart. The Center's director, Bob Sheets, briefed emergency managers to "enjoy their weekend" and check back in on Sunday (Sheets 1993b). By Saturday morning, however, Andrew had picked up speed, intensifying into a hurricane. That evening a Hurricane Watch was issued for the southeastern coast of Florida and at 8 a.m. Sunday it was upgraded to a Warning – about 21 hours before Andrew would make landfall (USDC 1993). Thus, the people of South Florida had a relatively short time in which to take this storm seriously. The early advisories were virtually ignored by many. In the 100 years or so for which records are available, South Florida has been the most hurricane-prone area of the continental United States. However, more than 30 years had passed since the last major hurricane, Donna, caused serious damage. In the interim, the region had grown rapidly and, as a consequence, most people living in South Florida in 1992 had never experienced a hurricane.

It became obvious on Sunday that Hurricane Andrew was going to make landfall somewhere along the southeast coast by Monday morning. Fortunately, it was a weekend when most people were at home, making it easier to get ready. Andrew strengthened and came through the Bahamas rated on the Saffir–Simpson Scale as a Category 4, nearly 5, hurricane. When a veteran of over 100 incursions into the eyes of hurricanes flew his *Storm Tracker* aircraft through Andrew on Sunday, he reported it as the roughest flight he had ever had (Historic Publications 1992: 18). Readings of 186 mph were recorded at a flight level of 10,000 feet (USDC 1993). By Sunday afternoon, it was undeniably clear that a dangerous storm was coming ashore in Florida somewhere between Palm Beach and the Keys. Hundreds of thousands of residents took to the roads to get supplies or to evacuate, creating the largest traffic jam the area has ever known. It is fair to say that most people, including local and state officials, were caught unprepared for a storm of this magnitude.

Sunday night was a restless one. It passed quietly, but in the early hours of

Monday morning, 24 August, the media began reporting Andrew's approach and most areas lost their electric service. At about 4 a.m. a radio report announced that leading winds had blown the radar from the roof of the National Hurricane Center in Coral Gables. Many, myself included, later described this as the moment they became truly alarmed. About an hour later the eye of the storm made landfall approximately 20 miles to the south of Coral Gables, near Homestead Air Force Base and the Leisure City area, continuing westward across the towns of Homestead and Florida City. Figure 1.1 depicts Hurricane Andrew's track and the pattern of the damage it left behind. Those at the storm's center were first hit with winds mostly from the north, followed by a very brief calm and second blast from the south, while the more populated areas just north of the eye were continuously battered (Wakimoto and Black 1994). Exact wind speeds continue to be debated, but Andrew is officially recorded as the third most intense hurricane to ever hit the continental United States, with sustained winds of at least 145 mph, gusting to at least 175 mph (Sheets 1993a).

In attempting to explain the damage patterns, experts collected evidence that this may have been an unusual hurricane with either vortices (Gore 1993) or tornadoes (Wakimoto and Black 1994) near or within its eye wall, and with some areas receiving higher second winds on the back side. It was a compact storm with hurricane-force winds confined to a relatively small area. As is common, it was asymmetrical – extending about 5 miles to the south and 15 miles to the north of the eye's path. The storm surge was estimated to have reached nearly 17 feet above normal sea level but, fortunately, it was limited to a short distance of shoreline. The rainfall associated with Andrew averaged between 4–5 inches, relatively little by sub-tropical standards (USDC 1993).

THE DAMAGE

The massive destruction inflicted on South Florida has been well documented.[2] It was not confined to a narrow path, but houses, cars, and community infrastructure were damaged throughout the Miami area. Debris blocked the streets in nearly every neighborhood. Life as we knew it had been severely disrupted. The numbers are impressive: 1.4 million without electricity; 150,000 homes without telephones; 3,300 miles of power lines destroyed; 9,500 traffic signs and signals out of order (Governor's Disaster Planning and Response Review Committee 1993). Everywhere, trees had been ripped up from their roots and whatever vegetation was left standing had been stripped of its leaves – our lush, green paradise had turned brown and barren overnight. The sights were so upsetting that people throughout the Miami area later reported assuming their neighborhoods had experienced the worse the storm had to offer. With no electricity, and thus no television, few people realized the extent of the tragedy which lay to the south.

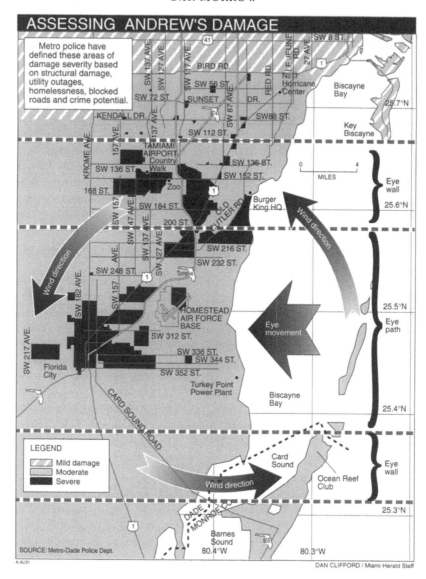

Figure 1.1 Assessing Hurricane Andrew's damage

Andrew had cut an 18 mile-wide path across the southern portion of Dade County – the area we will refer to as South Dade – creating what one journalist described as "a zone of destruction larger than the city of Chicago, or equal to 12 Manhattan Islands" (Gore 1993: 15). The infrastructure that had sustained a population of more than 375,000 had been virtually laid to waste. Nearly all public buildings, including 59 health facilities, 31 schools, and most fire and

Plate 1.1 A dazed man returned to his destroyed mobile home in Florida City to find everything he owned was gone except for this pair of pants
Source: C.M. Guerrero/*Miami Herald*

Plate 1.2 A resident of an apartment building lowers his belongings to the
street below
Source: Chuck Fadely/*Miami Herald*

police stations, were severely damaged or destroyed, along with about 8,000
businesses. Entire communities were literally wiped out. In the final analysis,
about 108,000 private homes were damaged, with about 49,000 of these
rendered uninhabitable (*Miami Herald* 1994, 24 August). More than 180,000
people were left homeless for some period of time (Governor's Disaster Planning
and Response Review Committee 1993). There had been 6,600 trailers or
mobile homes in South Dade (Metro Dade Planning Department 1992) and all
but nine were destroyed. Total losses have been placed in excess of $28 billion,
making this the costliest natural disaster in US history (Hebert, Jarrell and
Mayfield 1996). Yet experts warn that it could have been far worse.

The map in Figure 1.2 shows the location of the communities, as well as
neighborhoods, regions, and four bands we will be referring to throughout the
book. As you can see, if the storm had crossed 20 miles to the north, a distance
described by Bob Sheets as "a gnat's eyelash" in meteorological conditions and
too small to forecast even an hour before landfall, it would have passed over the
most populated area of Dade County (Sheets 1993a: 15). This includes the
islands of Miami Beach and Key Biscayne, the high rises of downtown Miami,
and dozens of communities, such as Liberty City, Hialeah, Miami Springs, and

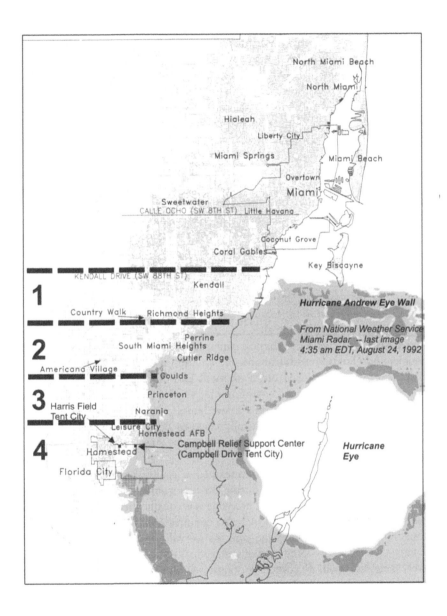

Figure 1.2 Bands of damage from Hurricane Andrew
Source: Hugh Gladwin/IPOR

Coral Gables – an area inhabited by about 1.6 million people and having property with tax values in excess of $60 billion (USDC 1993: 130). As stated by Kate Hale, then Dade County's Director of Emergency Management, "This was not the Big One of my nightmares" (Hale 1993). It was relatively small in area, moved quickly, carried little water, and hit the least-populated region of Dade County, home to only 18 per cent of the county's population.

In lectures around the country, Sheets emphasized that the damage would have been even worse if Andrew had hit outside of South Florida. While there were problems with Dade County's building codes necessitating that they be revised to deal with new building materials and construction technologies, at the time of Hurricane Andrew, Dade had the strongest wind-related building codes in the country. Code violations and shoddy construction practices have been highly touted as contributing to the damage, but it is now estimated that they accounted for less than 10 per cent of the losses (Sheets 1993a: 10). Considering the extent of the damage, the death toll was small – only 15 fatalities directly associated with the storm, nine of which occurred outdoors or in trailers, campers, or boats (USDC 1993). If Hurricane Andrew had hit elsewhere in the US with the same force, experts speculate that many more lives would have been lost.

THE VICTIMS

Our study goes beyond the physical damage caused by the storm to focus on the people, households, neighborhoods, and communities caught in its path. When Hurricane Andrew reduced much of South Florida to rubble, it destroyed the homes, jobs, institutions, and communities which sustained the lives of tens of thousands of households. Within a few hours, victims' previously taken-for-granted environments were drastically changed or completely lost, beginning with the roofs of their homes and extending outward for miles of nearly total destruction.[3]

No one in South Dade escaped unscathed. It could be argued that just trying to live and function in the area during the next couple of years was enough to qualify as an Andrew victim. Not everyone was equally affected, however. While the post-Andrew rhetoric often included comments that the storm was a "great equalizer," this was far from the case. As we will document, homes were not equally damaged. The same level of emergency and relief response did not reach every neighborhood. Recovery assistance was not equally distributed among those with similar needs. And, perhaps most important, individuals, households, neighborhoods, and even communities did not have the same recovery resources, either human or material.

An important focus of our research has been the extent to which victim attributes, such as social class, race/ethnicity, and gender, are associated with recovery progress and outcomes. At the time of the 1990 Census about half of the population of Dade County was composed of what traditionally are termed

Plate 1.3 A mother and daughter were able to rescue their cat but not much else
Source: Peter Andrew Bosh/*Miami Herald*

Plate 1.4 A man retrieves some of the family's belongings from their destroyed home
Source: C.W. Griffin/*Miami Herald*

minority households and the diversity within ethnic groups was interesting. As expressed by Portes and Stepick (1993: 8), "In Miami, the fragments of the [ethnic] mosaic are loose and do not come together in any familiar pattern." As shown in Table 1.1, the Hispanic component, roughly 49 per cent, is primarily Cuban, but also includes sizable proportions from throughout Latin America and the Caribbean. Similarly, while nearly 21 per cent of Dade Countians identified themselves as Black, only about half of these gave their origin as the United States. And, of course, the two classifications – Hispanic and Black – are not mutually exclusive, further defying simple categorization of South Florida ethnicity. For lack of a better term, the non-Hispanic, non-Black population is commonly referred to as Anglo, a label we use throughout this book. About 30 per cent of Dade County's 1990 population was Anglo. The area's households were also structurally diverse, with almost 15 per cent of the families headed by women, and it was not uncommon for non-nuclear kin to live together (Bureau of the Census 1992).

Table 1.1 Population of Dade County by country of origin

Self-identified ethnic group	Number	%	% of total
Hispanics			
Cuba	563,979	59.2	
Nicaragua	74,244	7.8	
Puerto Rica	72,827	7.6	
Colombia	53,582	5.6	
Dominica	23,475	2.5	
Mexico	23,112	2.4	
Honduras	18,102	1.9	
Peru	16,452	1.7	
Guatemala	8,242	0.9	
Ecuador	7,986	0.8	
Salvador	7,339	0.8	
Panama	6,729	0.7	
other	77,338	8.1	
Total Hispanics	953,407	100.0	49.2
Blacks			
United States	281,621	70.6	
Haiti	45,339	11.4	
Jamaica	27,204	6.8	
Bahamas	7,485	1.9	
other and uncoded	37,322	9.3	
Total Blacks	398,971	100.0	20.6
Anglos			
Non-Hispanic Whites	584,816		30.2
Total	1,937,194		100.0

Source: Metro Dade Planning, Development, and Regulation Department based on 1990 Census data

The hurricane's major impact area, as defined by the Metro Dade Planning Department (1992), included all of Dade County south of Kendall Drive (S.W. 88th Street). There were pockets of heavy damage north of this boundary, but most of the destruction occurred in the 270-square-mile area we refer to as South Dade. For purposes of description, Metro Dade divided South Dade into the four bands (illustrated in Figure 1.2). Population density declined as you moved southward – about 150,000 people lived in Band 1 near Kendall Drive, while Band 4 contained only 70,000 residents at the time of the hurricane. The area includes rural, suburban, and even some urban areas, high-rise office complexes and extensive farmlands, expensive homes and farmworker trailer camps. While there are some well-defined suburban communities, such as Country Walk and the retirement community of Americana Village,[4] most of South Dade can best be described as a vast area of scattered, ill-defined neighborhoods, shopping strips and centers, pockets of industrial buildings and warehouses, and undeveloped agricultural areas.

Ethnic diversity in South Dade prior to Andrew was somewhat different from the county as a whole. With the population consisting of about 30 per cent Hispanics and 18 per cent Blacks, compared to 49 per cent and 21 per cent countywide, the area included some of the more predominantly Anglo neighborhoods in Dade County. Blacks were concentrated in older communities such as Perrine, Goulds, Florida City, and the suburban development of Richmond Heights. Hispanics, while less segregated, tended to reside in the more affluent suburbs near Kendall Drive, in the working class community of South Miami Heights, and in the southernmost band where sizable concentrations of Mexicans and Haitians lived, including many agricultural workers.

Most of the Greater Miami area is unincorporated, and governed by Metropolitan Dade County. At the time of the storm, there were only two incorporated towns in South Dade – Homestead and Florida City – and they differed dramatically. In 1990, Homestead had about 27,000 residents, compared to less than 6,000 in Florida City. Historically, Homestead was generally thought of as a town of middle class and well-to-do Anglo farmers, while Florida City was the poor Black community across the tracks. In truth, according to 1990 Census data, most of the residents of both towns were renters with modest or low incomes. Homestead Air Force Base was a vital part of the economy and social environment of the entire area. It is estimated that the base's destruction by Andrew caused the disappearance of 8,000 jobs and $405 million in the local economy (Office of County Manager 1994).

In 1990, there were 130,000 dwellings in South Dade. Most were owner-occupied, single-family homes, but there were multi-family units scattered throughout the area, with low-rent housing heavily concentrated in the lower band. Virtually all of the public housing units in South Dade – over 1,600 federally funded and 5,500 state funded units – were destroyed by the hurricane, along with most subsidized (Section 8) rental housing (Metro Dade 1994b). The households of South Dade, compared to county averages, had a higher proportion of family units (about 76 per cent), were slightly larger (2.96 people), and had more children under age 18 (23 per cent). While only 8.4 per cent of the population was over 65 years old, it was concentrated in some of the worse hit areas, such as Naranja and Goulds. The elderly population was disproportionately poor, with about 73 per cent relying solely on Social Security income (Atkins, *et al.* 1993). The median household income in South Dade of about $38,000 belies the diversity. The more affluent areas, such as Gables by the Sea, West Kendall, the Hammocks, and Country Walk, are located in the northern portion of the impacted area. Moving southward, socioeconomic status tended to decrease, as illustrated by an increase in the proportion of households with incomes under $15,000 – from 11 per cent in Band 1 to 29 per cent in Band 4 (Metro Dade Planning Department 1992). In general, it can be said that the heterogeneity of the victims of Andrew, in terms of class, ethnicity, and household structure, is in line with projected trends for the United States as a whole (O'Hare 1992; Ahlburg and DeVita 1992). This makes South Florida a particularly exciting and relevant setting for disaster research.

THE RESEARCH TEAM

Within a week after the storm a disaster team of faculty and graduate students from Florida International University (FIU) was assembled by Walter Peacock (Associate Professor of Sociology and Research Director of the International Hurricane Center), with initial support from President Modesto Maidique and the Division of Sponsored Research. Our team of sociologists and anthropologists included Elaine Enarson (Adjunct Professor of Sociology), Chris Girard (Associate Professor of Sociology), Hugh Gladwin (Associate Professor of Anthropology and Director of FIU's Institute for Public Opinion Research), Guillermo Grenier (Associate Professor of Sociology and Director of the Center for Labor Research and Studies), Barry Levine (Professor of Sociology), Betty Morrow (Associate Professor of Sociology), and Kevin Yelvington (Assistant Professor of Anthropology). It profited from the contributions of several dedicated graduate students, including Nicole Dash, Linda Beer, Manny Alba, and Donna Kerner. The entire manuscript benefited from the expert editing skills of Kathleen Ragsdale. While our interests, expertise, and preferred research methodologies varied, we were united by a common interest in studying the effects of Hurricane Andrew on the households and communities of the area.

In a very real sense, team members were both researchers and victims. Most of our homes were damaged, some to a considerable degree – Walt's family was without a permanent residence for months, and it was nearly a year before Elaine returned to her home. The University's main campus sustained considerable damage, delaying the opening of fall classes. Nevertheless, we were soon in the field, beginning the first of eight major projects which form the basis for this book. We begin with a brief overview of each. (Further details about funding sources, sampling, data collection, and methodologies are provided in the appendix.)

OUR PROJECTS

Tent city study

We began by studying the experiences of some of Andrew's most severely impacted victims – residents of the tent cities established by the federal government, for displaced victims with no other housing alternatives. We conducted over fifty open-ended interviews with personnel from the US Army and Marine Corp, the American Red Cross (ARC), the Federal Emergency Management Agency (FEMA), and, most importantly, homeless individuals and families. This qualitative project provided important insights into the circumstances of severely impacted victims including their experiences prior to, during, and after the storm, helping us to identify problems and issues for subsequent research activities.

FIU Hurricane Andrew survey

While the tent city interviews provided detailed information about an important group of victims, we wanted a more representative picture of the impact of Andrew on South Florida households. We were also interested in learning about household preparation and evacuation activities while the memories were still relatively fresh. Funded by a grant from the National Science Foundation, about four months after the storm we conducted a telephone survey of a random sample of over 1,000 households in Dade County. In spite of the damage sustained to the infrastructure, a telephone survey was a viable option since service had been restored in most areas. The telephone company had also instituted call-forwarding and recorded message services, allowing us to track down many dislocated or moved households. Utilizing supplemental funding from the John S. and James L. Knight Foundation's sponsorship of FIU's *Lessons Learned from Hurricane Andrew* conference, an additional over-sample of 300 South Dade households was undertaken to better insure coverage of the most heavily impacted area. Respondents were asked about household preparation and evacuation activities, dislocation and relocation, household damage, insurance settlements, and other sources of assistance.

South Miami Heights survey

To provide more in-depth insight into the process of household recovery, our next project focused on one heavily impacted neighborhood, South Miami Heights. This working class, culturally diverse community sustained major damage, yet was virtually ignored by authorities and agencies in the immediate

13

aftermath. After remapping the community, a random sample was drawn and interviews conducted at approximately 200 South Miami homes. Our interview schedule included questions about damage, sources and amounts of assistance, insurance, contractors, community recovery, persistent family and community problems, and the impact of this experience on household members. Observations made while canvassing the neighborhoods and visiting homes were also a valuable source of information.

Family impact study

Andrew destroyed homes, not just houses. The disruption in South Dade took many different forms, including household and job loss or dislocation, extended commuting patterns, living in crowded and often badly damaged structures, dealing with the maze of paperwork and tasks associated with loss recovery and household reconstruction, as well as the lack of community infrastructure, such as parks and recreation facilities, neighborhood stores, and local services. To address the effects of these daily hardships on families, Enarson and Morrow conducted open-ended interviews with care providers, counselors, school and church personnel, and women's groups. Focus groups with single mothers in a public housing project, low-income Haitian women, family day care providers, and battered women were also held. The goal was to understand better the dynamics of family response to disaster, with special emphasis on groups appearing to be having great difficulty. Since women continue to be the primary homemakers in most families, they were likely to be in the best position to reflect on the household effects of a disaster. Input from these *experts* is seldom solicited by community leaders and those responsible for making policy and resource decisions that directly impact on households.

South Dade population impact study

Population change can be seen as either a cause or an effect of long-term community recovery, depending upon your perspective. People will return if the economy comes back, but the economy is not likely to recover if people do not return. In addition, state funding and revenue sharing are typically based on the size of the population of an impacted area. Thus, demographics can dramatically impact the resources allocated for long-term recovery. Each year the Bureau of Economic and Business Research (BEBR) at the University of Florida derives estimates of statewide population changes. Tax assessments and building permit records are used to estimate the stock of housing which, combined with information on average household size, is then used to project US Census estimates to the current year. The utilization of this standard methodology in post-hurricane South Dade, however, would have resulted in a major undercount since the housing stock had been drastically reduced and the size of many households altered. Furthermore, higher rental densities and more people living in

Plate 1.5 Migrant workers head back to their damaged dwellings with food and clothing from the Haitian Mission United Methodist Church
Source: Carl Juste/*Miami Herald*

non-standard housing arrangements were anticipated. For these reasons, BEBR sub-contracted with us to provide a more accurate estimate of Dade's population. Large sections of South Dade were remapped and nearly 3,000 field inspections and interviews were conducted in inhabited housing to determine occupancy rates, average household size, how many were living in non-standard arrangements, and to provide information on post-Andrew movement and insurance settlements. This more accurate picture of population change was then incorporated into BEBR's 1993 estimates.

American Red Cross project

In a study funded by the American Red Cross (ARC), its unprecedented Hurricane Andrew Recovery Project was evaluated as part of an organizational effort to determine the most effective ARC long-term recovery function after a high-impact disaster. Through in-depth interviews with about fifty key informants from the ARC and other agencies involved in the recovery effort, Morrow, Enarson, and Peacock identified several areas in which agency resources and expertise were especially needed to promote long-term recovery. ARC's special long-term client intervention project met with limited success and was the subject of considerable intra- as well as inter-agency controversy. We

15

recommended that ARC consider providing long-term assistance indirectly through other agencies working in a stricken area, focusing its attention on providing leadership and training in case management. The organization's unique relationship with FEMA places it in a central position to serve as a clearinghouse for client information. While the major product of this project was an internal report for the ARC, valuable information about victims and providers enriches several chapters of the book.

Homestead housing needs and demographic study

Hurricane Andrew altered Homestead in dramatic fashion. According to BEBR's estimates, by 1 April 1993 its population had dropped by over 30 per cent. Therefore, 1990 Census data were nearly useless to town officials trying to determine the needs of post-Andrew residents, particularly related to housing. The only viable solution was to conduct a more thorough study, utilizing sampling procedures to obtain population estimates and demographic information. With Thomas Wilson of the FIU/Florida Atlantic University Joint Center for Environmental and Urban Problems, a randomly selected sample of approximately 1,000 households were surveyed by telephone regarding housing characteristics, such as adequacy, costs, and state of repair, as well as demographic information about household members.

Plate 1.6 Hand-painted messages throughout South Dade communicated useful information, prayers, and poignant appeals, as well as humor
Source: Charles Trainor, Jr./*Miami Herald*

Florida City study

From the onset, it was clear that South Dade's other incorporated community, Florida City, was having a difficult time with recovery. Utilizing tax and business data, supplemented by information from a variety of sources, Nicole Dash studied this community as a thesis project. In her analysis, she compared Florida City to its neighbor of Homestead in terms of hurricane impact, as well as business and economic recovery.

Emergency management organizational analysis

We are pleased to be able to supplement the work of our disaster team with an analysis of early governmental response by two of our colleagues at Florida International University – Harvey Averch (Professor of Public Administration) and Milan Dluhy (Professor of Public Administration and Director of the Institute of Government). In order to analyze intergovernmental decision making and response during the emergency phase of this disaster, they interviewed key decision makers at different levels of government.

Our work has been a cooperative endeavor in which we have all shared. While we have taken turns authoring various parts of the book, we have collaborated on nearly all research projects, data analysis, and writing tasks. In a similar fashion, there is no clear delineation regarding which project is reported in which chapter; rather, we set out to learn whatever we could, using whatever methodology seemed most appropriate to the task, and we have combined the findings from the various projects around several themes. The results reflect the eclecticism of our personal styles and backgrounds, as well as that of our subjects.

ORGANIZATION OF THE BOOK

From the onset our work was guided by the perspective that the process of responding to a natural disaster should be conceptualized as an inherently social one occurring within a broader context – a socio-political ecological field, as we prefer to think of it. In Chapter 2, Walt Peacock, with Kathleen Ragsdale, discusses this theoretical perspective and argues that it is within an inherently competitive and conflictual atmosphere that individuals, families, and communities, compete for the resources to respond to a disaster.

Guillermo Grenier and I elaborate on the socio-political context of Greater Miami, with its unique pattern of racial, ethnic, class, and gender relations. Experiencing a disaster of the magnitude of Hurricane Andrew in a metropolitan area with the diversity and dynamism of Miami provides a truly unique opportunity to study the sociology of disaster. Of particular relevance are ways in which economic and political conditions predisposed certain segments of the community to be disproportionately impacted and placed them at a disadvantage during the competitive recovery period.

17

Our Andrew story begins with Gladwin and Peacock describing events of the hectic weekend preceding Hurricane Andrew. In Chapter 4, they utilize information from various field and telephone interviews to discuss how South Floridians received information about the storm's progress and what they should do, how they prepared their homes, and who evacuated. Throughout this project it has been important to us that our work should have practical application, and the chapter ends with a discussion of several policy implications. In the following chapter Averch and Dluhy provide an organizational analysis of emergency response during the first six days of the crisis.

The next three chapters focus on the experiences of the victims. Yelvington discusses life in the tent cities and issues surrounding their establishment, management, and closure. Enarson and I focus on the crucial, yet typically overlooked, roles of women at all levels of community and household response. In our qualitative analysis, we develop several composite profiles to represent the experiences of the women we interviewed. In Chapter 8, I draw from a variety of sources to discuss the families of Hurricane Andrew, including the extent to which kin networks were important, the effects of family attributes on response and recovery, and the internal effects of this experience on family life.

In Chapters 9 and 10, Girard and Peacock report findings from several survey projects on the experiences of Andrew's victims related to housing damage, insurance, and relocation, with particular focus on the consequences of race and ethnicity on the outcomes. Along this same line, Dash, with Peacock and Morrow, presents her work on the disadvantaged position of the small Black community of Florida City throughout this entire process. She provides ample evidence of how preexisting economic and political conditions at the community level influence the effects of a disaster on its residents. In the final chapter, we use examples from these studies to illustrate how natural disasters initiate community changes, including adaptations in institutional responses, inter-organizational relationships, stratification, and population patterns. Many changes have occurred as a result of this experience. Three years later Dade County, particularly South Dade, is a different place in some respects, which we will discuss, but it is also unchanged in many ways, including its persistent patterns of racial, ethnic, and gender stratification.

NOTES

1 Often quoted from Charles Schultz' *Peanuts* cartoon, this was the opening line in the 1832 novel *Paul Clifford* by Edward George Buliver-Lytton.
2 Several documentary books provide details about Hurricane Andrew and its effects on the environment and people of South Florida, such as *The Big One – Hurricane Andrew* published by the *Miami Herald* (1992a), *Andrew! Savagery from the Sea* by the *Sun-Sentinel* in Fort Lauderdale (1992), and *Hurricane Andrew: Path of Destruction* by Historic Publications. A local television station, WTVJ, compiled an excellent documentary video, *Hurricane Andrew: As It Happened* (1992).

3 For an ethnographic account of life in "the zone," see Smith and Belgrave, "The Reconstruction of Everyday Life: Experiencing Hurricane Andrew" (1995).

4 For more information on strategies used by the residents of this retirement village of prefabricated homes to manage their own community recovery, see Guillette, *The Role of the Aged in Community Recovery Following Hurricane Andrew* (1993).

2

SOCIAL SYSTEMS, ECOLOGICAL NETWORKS AND DISASTERS

Toward a socio-political ecology of disasters

Walter Gillis Peacock with A. Kathleen Ragsdale

For residents of southern sections of the Miami metropolitan area, the morning of 24 August 1992 represented a dramatic discontinuity in the world as they knew it. At least temporarily, there were no daily routines or normal lives. Everyday activities, such as getting a drink of water, fixing a meal, or taking a bath, became daunting tasks. The information upon which the routines of daily life had been structured was virtually useless because the "stuff" needed to put it to use was no longer available – no electricity, water, gas, telephone, refrigerator, stove, bed, bathtub, room, or for many, even house. There were no stores or businesses to buy food, supplies, or services. Schools, hospitals, recreation facilities, entire neighborhoods, were virtually gone. The world was now full of uncertainty; new information was needed – where to find things, how to fix things, and how to get things done.

As the hours turned to days, the level of uncertainty worsened for many. There seemed to be no organized activity dealing with the crisis. The pervasive level of frustration was captured by the questions, the choir of questions, heard at every turn. Where do we go for help? Where will we live? How can I protect what's left? Will my insurance cover this? How long will it take to rebuild? Will life ever return to normal? Who's in charge? Does anyone care?

People wanted to be able to point to one place, one person, or one organization for simple, straightforward answers. What became increasingly clear was that no single person or organization was in charge, and there would be no simple solutions. Eventually, a multitude of suggestions and ideas emerged, some of which worked while others did not. Some were novel solutions, others were "old hat." Some resulted in new ways of thinking about communities, enhanced mitigation, and increased community involvement, while others reduced involvement, created dependency, and increased future vulnerability. For parts of the community, typically areas lacking local resources, few or no solutions were even proffered.

20

In this book we focus on some of the important questions raised and examine some of the solutions offered as families, households, businesses, and other community groups and organizations prepared for the impending storm, made evacuation decisions, dealt with the immediate crisis, and negotiated the lengthy restoration and recovery processes. We explore the discourse associated with a mosaic of solutions, seeking to understand why some paths were taken over others and examining some of the consequences. And, we ask why some community issues were never raised, or received scant attention, at least from those with the power and resources to address them.

In a sense, we, too, were searching for what could not be found – for there were no simple solutions and there are no simple findings. And yet, some themes clearly emerged out of Hurricane Andrew's chaos. They emerged, not from the storm itself, but rather from the complexity of preexisting community structures. While disaster agents, such as hurricanes, may appear to be simple physical events, in order to understand their impact, including how communities respond and are ultimately affected, we must understand the nature of community itself.

THE NATURE OF HUMAN COMMUNITIES

When contemplating what is meant by the term community, ideas such as "a common sense of identity," feelings of affinity and cooperation come to mind, particularly in connection with our romantic notion of small rural community. Let the focus shift to life in today's cities, however, and we are more likely to think of crime, conflict, competition, and exploitation. These seemingly antithetical positions are evident, not only in everyday discourse, but also in the social science literature. The position adopted in our work is that any community, regardless of its size, location, or level of development, should be conceptualized as an *ecological network* of social systems. This network, along with the bio-physical characteristics of the more or less identifiable physical space it occupies, form an *ecological field* (Bates and Pelanda 1994).

When considering network process and operation, we adopt a *socio-political ecology* perspective. This position draws upon emerging currents in the study of disasters calling for a broader ecological approach, and goes beyond examining the interaction between social systems and their bio-physical environments, focusing on interactions among social systems themselves (for example, Hewitt 1983; Faupel 1985; Faupel 1987; Pelanda 1989; Bates 1993; Bates and Pelanda 1994). The notion of studying the socio-political ecology surrounding an event is consistent with increased interest in political economy and critical perspectives, including the analysis of minority, gender, and inequality issues at all phases of disaster research, from evacuation through mitigation (Perry and Mushkatel 1986; Bolin and Bolton 1986; Scanlon 1988; Tierney 1989; Phillips 1993b; Morrow and Enarson 1994; Morrow and Enarson 1996). All these properties reflect an awareness of conflict and competition and demand attention to the complexity, heterogeneity, inequality, and contingency found in human communities if we are to understand social processes and structures, including those associated with disasters.

COMMUNITY AS AN ECOLOGICAL NETWORK

Ecological approaches have a long history in the study of human communities (Park 1936; Duncan 1961; Duncan 1964; Odum 1971) and interest has risen considerably since the mid-1970s (Catton and Dunlap 1978a; Catton and Dunlap 1978b; Dunlap and Catton 1979; Boulding 1981; Hannan and Freeman 1989; Laska 1993). This trend is evident in recent disaster research and theory suggesting the utility of central concepts, such as an *ecological complex* (Faupel 1985 and 1987). While bemoaning an overemphasis on the interaction between human systems and their environment, this work suggests it is important to examine the interaction among cultural and organizational components of human communities themselves. The position we take extends this theme, directly addressing the nature of human communities and paying particular attention to conflict and competition within them.

We begin by rejecting the notion of community as a single, bounded, autonomous social system and suggest that it be conceptualized instead as an ecological network of interacting social systems.[1] This demands that an ecological framework be applied, not only to the relationship between a social system and its bio-physical environment, but also to the relationships among the various social units comprising the network itself (Pelanda 1989; Hannan and Freeman 1989; Peacock 1991; Bates 1993; Bates and Pelanda 1994). Groups and organizations which are semi-autonomous self-referential social systems (Luhmann 1990a and 1990b) are linked together through relationships and member-sharing to form an ecological network of contingent relationships.

The critical factor in conceptualizing community as an ecological network stems from the nature of the relationships among the constituent systems (Bates 1993; Bates and Pelanda 1994). These linkages are not bonded in the sense of there being rules (norms and roles) designating particular linkages, as is often the case within organizations. For example, most households are not linked in a particularistic fashion to specific retail businesses (such as particular grocery, hardware, or drug stores). Whether a household is linked to a particular business depends on factors like convenience, price, quality, information, and history as determinants in this inherently stochastic process (Bateson 1972).

When this idea is expanded to consider the range of interactions among the multitude of units making up a community, it becomes clear that the aggregation cannot be thought of as a single autonomous social organization. The units are not bound by systemic, bonded, unit-specific linkages implying regulation, coordination, and central control. Rather, the groups and organizations within a community are linked by sets of contingency linkages among which information, members, and resources flow. Any coordination or control is not likely to be the result of a centralized authority structure, but rather to have emerged out of the interplay of mutual contingencies, competing interests, and coalitions exercised through a variety of structural linkages (Bates and Harvey 1975; Peacock 1991). Indeed, coordination and conflict resolution become critical to network processes.

Rejecting the idea of community as a single system does not dispense with the notions of dependence and interdependence. An ecological network is structured by a division of labor in which social units occupy niches in a web of interdependencies (Bates 1993: 250). Units making up a social network may be highly specialized, producing few of the resources needed to carry out their own activities. As a result, they will be dependent on categories or sets of other systems. Thus, dependency and interdependency are fundamental characteristics of a community and have important implications for power, influence, and survival within the ecological network.

While groups and organizations are conceptualized as self-referential systems having autonomous authority structures, this does not imply equality. Some exert far greater influence than others. Power and resource distribution are critical determinants of network operations and system survival. The outcome of competition and conflict over the control of resources will determine relationship patterns, as well as the long-run survival of specific social units within the network. Furthermore, these factors will, to a large extent, determine the interaction between the social network and its larger ecological field. For example, from a mitigation viewpoint, business and development concerns overly influence land use policy and building patterns (Logan and Molotch 1987). In policy decisions, profits, not field sustainability nor the ability of the built environment to withstand potentially hazardous environmental impacts, typically take precedence.

The historical development of government can be viewed as an evolution of systems coordinating the relationships among the various groups and organizations within ecological networks (Braudel 1981; Braudel 1982; Braudel 1984; Lenski and Lenski 1990). The degree of government control, in part, determines the extent to which a community approximates a single system. Communities in Western democracies tend to be loosely fitting networks, especially when compared to the more centrally planned communities and societies of totalitarian regimes. In addition to their coordinating activities, governmental organizations serve as conflict resolution structures (Bates and Bacon 1972). Competing interests resolve conflicts in a more or less controlled fashion through legislatures and courts, although access to and success in these structures certainly varies. The legislative process sets the rules for exchange and declares certain exchanges as non-contingent and outside the market (Bates 1974). Executive branches administer non-market service programs and attempt to regulate, coordinate, and police activities within the network. Taken as a whole, government represents multiple organizations and agencies, themselves constituting a sub-network (Bates 1993). Competition and conflict over jurisdictional issues among national, local, and state governments adds further complexity to local community networks. Ironically, conflict, competition, and lack of coordination are inherent among these very organizations and agencies expected to coordinate and regulate a community's ecological network.

A host of non-governmental coordinating bodies exist in the interstices between community groups and organizations (Bates and Harvey 1975). Business coalitions, better business bureaus, unions, and intergovernmental committees, as examples, coordinate actions and relationships throughout the network and influence future events, set prices, control policies, and selectively stimulate or thwart activities related to their collective interests. They often link government and business interests, thereby insuring that regulations, policies, and agencies' actions do not negatively impact business concerns. Some of these groups, such as the interlocking directorates and informal coalitions of power elites, fall outside public knowledge, either by design or due to network complexity. Certain organizations come to dominate particular niches, exerting inordinate control over network activities. While these factors insure that social entities are not playing on equal fields, they do lessen the contingencies and uncertainties of network operation.

As a perspective then, socio-political ecology extends an ecological perspective to the very heart of what Park (1936) referred to as the organizational component of the ecological complex – rejecting the notion of community as a single autonomous social system. Rather, a community is an ecological network of groups and organizations linked through divisions of labor based on contingent relationships. Competition and conflict are inherent, hence mechanisms of conflict resolution and coordination are crucial to the long-term functioning of the network. Finally, differential access to network resources is critical for understanding the survival and reproduction of its social units. The implications of this perspective, particularly these latter points, will become clear as we turn our attention to a discussion of disaster and recovery, focusing on the processes of household recovery in the United States.

DISASTER AND RECOVERY

A major natural disaster, in the sociological sense, can be thought of as a failure of the social systems constituting a community to adapt to an environmental event (Pelanda 1982; Bates 1982; Britton 1986; Kreps 1989; Pelanda 1989). This failure is not simply the result of a high-impact natural phenomenon, such as a hurricane or earthquake. It should also be viewed as the failure to develop and distribute, among other things, technology in the form of housing and community infrastructure capable of withstanding such an event. The disruption of networks of social interaction and the inability of social actors to operate on the basis of normative information are what define the disaster as a social event. Information relevant for normal behavioral patterns now lacks utility because stores, buildings, transportation, and energy systems are typically unavailable.

Restoration of these facilities is usually beyond a household's limited capacities and is dependent on parallel restoration processes occurring among units throughout the network. Community recovery from this perspective can be thought of as a process in which the groups and organizations making up the community attempt to re-establish social networks to carry out the routines of daily life. In terms of linkages among the units, recovery demands that the contingencies for obtaining the necessary knowledge, goods, and services within the social network be brought within acceptable parameters.

A critical component of recovery is the reaccumulation of the capital or physical infrastructure used by the various social units making up the network. For example, households use many items such as stoves, refrigerators, and washing machines to carry out daily routines. Thus, an important measurement of household recovery is the reaccumulation of these material items, thereby allowing routine patterns to re-emerge (Bates and Peacock 1993). Household recovery is a dynamic process where households, as interdependent social units, interact with their environments to re-establish their living conditions and patterns of interaction. While cooperation certainly exists, recovery typically entails sets of negotiations that can best be characterized as competitive, potentially conflict-ridden, and stressful. Many of the network's social units occupy similar niches, placing them in competition for scarce resources and services. This is particularly the case with households which, as end consumers, must enter into a complex, competitive ecological network *en masse* in order to negotiate their own recovery. Many factors related to community complexity have consequences for the outcome of this recovery process, such as its level of economic development, size, division of labor, and political system (Hoover and Bates 1985; Peacock, Killian and Bates 1987).

A single recovery pattern rarely emerges from the complexity of processes and factors influencing long-term recovery. The outcome for any particular household is contingent upon a host of factors and is likely to be highly uneven across households. The results may include improved living conditions, enhanced ability to withstand future disasters, even economic improvement and increased development (Senior 1970; Bates, Farrell and Glittenberg 1979; Bates and Killian 1981; Catarinussi, Pelanda and Moretti 1981; Abril-Ojeda 1982; Geipel 1982). On the other hand, decline in socioeconomic status, failure to regain predisaster living conditions, and increased vulnerability may occur (Clifford 1956; Bolin 1982; Bolin and Bolton 1983; Peacock, Killian and Bates 1987; Bates and Peacock 1987; Bates 1982).

The socio-political ecology perspective is concerned with how social structures shape the dynamics of household recovery, resulting in these disparate outcomes. Relevant factors include household attributes and access to financial, medical, material, and informational resources. The policies and programs of governmental agencies, the response of private organizations and businesses (often captured under the rubric of the market's response), and the role of non-profit non-governmental agencies and local development agencies play

25

significant roles in determining and altering the resources available for household recovery (Mileti, Drabek and Haas 1975; Bates 1982; Bolin 1982; Kreps 1984; Drabek 1986; Bates and Peacock 1987; Bates and Peacock 1989b). Thus, varying patterns of recovery or non-recovery from a disaster emerge as individual households with differing attributes negotiate the contingencies of their unique ecological network in order to mobilize vital resources and services.

DISASTER RECOVERY IN THE UNITED STATES

Unlike some Western industrialized nations where the government takes a direct role in planning, financing, and implementing post-disaster reconstruction, the United States relies heavily on private insurance payments, supplemented by government-sponsored low interest loans and grants, with non-profit voluntary agencies often filling in some of the gaps. It is essentially a market-based recovery approach. In high-impact disasters, local and regional resources are insufficient and must be supplemented from outside the community. These supra-local resources are generally filtered though existing organizations (for example financial institutions, insurance companies, governmental and non-governmental agencies). Disaster-related federal and state assistance programs generally operate under constrained mandates which set qualification standards and, until very recently, have narrowly defined restoration goals. In the final analysis, it is up to individual households to negotiate the relevant processes to acquire the necessary funds – whether insurance settlements, government or private loans or grants, or their own resources – to use in the market place to facilitate their own recovery.

It is a general, though not fully tested, contention that a market-based recovery mode is inherently conservative in nature. Simple restoration, rather than equity and development, is the intended outcome (Bolin 1982; Bates and Peacock 1987). Preimpact failures of market and regulatory mechanisms, as well as pre-existing social inequities, find full play during recovery and reconstruction. For example, engineers have noted that disaster damage is, to a large extent, preventable if appropriate building technologies, in terms of designs, materials, and construction techniques, are utilized. However, in today's mass housing markets, construction techniques, as well as government regulatory mechanisms, such as building codes and inspections, often fail to insure adequate and appropriate housing. These matters typically receive a great deal of attention immediately following a disaster, but the urgency of citizens to rebuild, coupled with political pressure from developers and builders, tend to thwart reform. As a result, many pre-impact failures are reincorporated into the rebuilt environment.

Determinants of household attainment and accumulation, such as socio-economic status and household composition, take on added significance after a disaster (Cochrane 1975; Haas, Kates, and Bowden 1977; Bolin 1982; Drabek 1986; Peacock, Killian and Bates 1987; Bates and Peacock 1993). Not only do high-income households have more personal reserves to draw upon, they may also receive more federal disaster assistance (Bolin 1982). Household composition is important because resources and needs vary according to number of adults, particularly wage-earners, as well as the gender and age distribution. Its stage in the family life cycle can also affect economic, as well as emotional, family recovery (Bolin and Trainer 1978; Bolin 1982). The extent to which households are connected to the community, in terms of family, ethnic, and other social networks, is an important factor influencing disaster response and outcome (Quarantelli 1960; Hill and Hansen 1962; Drabek and Boggs 1968; Drabek et al. 1975; Drabek and Key 1976; Drabek and Key 1982; Bolin 1982; Perry and Mushkatel 1986; Portes and Sensenbrenner 1993).

Race and ethnicity can have important recovery consequences, a factor requiring increased consideration given the changing population demographics of the United States. Minority households tend to have significantly lower incomes (O'Hare 1992). Discrimination and cultural factors can further limit their access to important public and private resources, such as loans and adequate insurance settlements. Market-based recovery policies tend to magnify the consequences of these conditions, placing minority households at much greater risk of failing to recover. And yet, local community situations may modify these findings. As we discuss in the next chapter, the case of Cuban Americans in South Florida represents a situation where the creation of a comprehensive sub-network has enabled a national minority largely to avoid disadvantaged status. Specifically, the establishment of an enclave economy has allowed Cubans to move quickly into the political and economic hierarchy of the region, if not the nation (Wilson and Portes 1980; Portes and Bach 1985; Pérez 1992; Portes and Stepick 1993). As a result, Cubans tend to enjoy greater access to resources than do most minority groups (Grenier and Stepick 1992). One of the factors we will subsequently address is the extent to which this enclave may have facilitated the recovery of Cuban households after Hurricane Andrew.

SOCIO-POLITICAL ECOLOGY AND DISASTER RESEARCH

Applying an ecological perspective to the study of disaster brings to the forefront socio-political issues such as the extent to which social inequality, heterogeneity and complexity, competition and conflict, and coordination exist within the network of social systems. The following sections discuss and provide examples of ways in which these factors influence disaster response and outcomes, advocating their emphasis in future disaster research. This discussion is not meant to be exhaustive and may well bring to mind additional issues associated with the socio-political ecology of a disaster setting. Our intent is to illustrate the types of issues arising from this conceptual framework, many of which are addressed in our work.

Social inequality

It is important that we continue to study the effects of political and socio-economic inequality – whether associated with race, gender, age, class, or some other attribute – on all phases of disaster processes. The consequences of various types of social differences for the modes and results of recovery have begun to receive attention, but many related issues have yet to be fully identified, conceptualized, and studied. This is particularly the case with the role and consequences of gender in disaster events, from household preparation and evacuation to long-term recovery efforts (Morrow and Enarson 1994). Due to the paucity of gender-related work, in spite of its theoretical significance in all levels of social phenomenon, gender is a central theme in our work. Enarson and Morrow (Chapter 7) offer a gendered analysis of disaster focusing on women's experiences after Hurricane Andrew. They are particularly concerned with ways in which gender intersects with race and class to place women and their households at special risk. Morrow then highlights women's perspectives in an analysis of overall family response (Chapter 8). She also examines ways in which internal characteristics of families and households influenced the effects and results of this experience.

South Florida's population provided unique opportunities for examining the influence of diversity beyond the household as well. Yelvington addresses some subtle issues of race and ethnic relations as households living in temporary tent cities struggled to cope with the initial aftermath (Chapter 6). Gladwin and Peacock address racial differences in preparation and evacuation decisions (Chapter 4). While recent research suggests few, if any, consequences, their unique analyses and findings suggest that normal modeling of evacuation decisions obscures many of the differences. The consequences of racial and ethnic inequality in access to insurance are addressed by Peacock and Girard, thereby assessing some of the consequences of market-based recovery policies (Chapter 9). Girard and Peacock focus on an issue that has received scant attention – the demographics of disaster – while assessing ethnic and racial differences in post-disaster population movement (Chapter 10). And Dash, Peacock and Morrow examine the restoration progress of a predominantly Black community in the hardest hit part of South Dade (Chapter 11).

Inequality issues associated with political participation, representation, and power in populations impacted by disasters have not received the attention they warrant. However, over the last two decades events in Nicaragua, Guatemala, and Mexico have begun to sensitize the field to the linkage between disaster and political change. Following the 1976 earthquake, recovery and development programs in Guatemala set into motion political change that greatly improved many communities – until the military's "scorched earth" policy brought these reforms to a bloody and brutal end (Bates and Peacock 1987; Peacock 1994). Similarly, the Mexico City earthquake of 1986 became a visible symbol of the problems of Mexico's ruling political party (PRI) and became a watershed event in the emergence of credible grass-roots opposition movements (Castanos-Lomnitz 1993; Poniatowska 1995). Research conducted in the West sometimes fails to consider the importance of these factors by assuming democracy and political participation to be the rule.

Hurricane Andrew struck less than three months before the 1992 presidential elections. It has been suggested that the slowness of the initial federal response made President Bush appear ineffectual, and hurt his re-election bid. A key FEMA administrator recounted how, subsequent to the President's visit to devastated South Dade, they received orders to quickly get out as much assistance money as possible. Yet he questioned the wisdom of distributing Individual and Family Grants before stores had reopened in the community where the replacement household belongings could be purchased. In another example, during the first few months after the storm, most predominately Black neighborhoods and unincorporated areas received scant attention from authorities. It so happened that the November 1992 election was the first to be held after Florida electoral reform had resulted in redrawn district lines insuring better minority representation at all levels of government (Morrow and Peacock 1993). After several African-American officials were subsequently elected at local, state, and federal levels, minority issues gained increased attention, as illustrated by the appointment of several special governmental commissions and officials to oversee the recovery of neglected areas. It is difficult to know the extent to which population changes prompted by Hurricane Andrew played a role in the elections, but the improved minority representation no doubt contributed to a re-adjustment of governmental and private sector response.

The extraordinary influence exerted by powerful economic interests on government policy, land use patterns, and construction has important implications for disaster research. As previously discussed, pro-development interests can dramatically alter attempts to regulate development, financial decisions, and building practices (Logan and Molotch 1987). One of the most insightful examinations of this influence in legislative and regulatory processes was undertaken following Hurricane Andrew, not by social scientists, but by reporters from the local newspaper (*Miami Herald* 1992n, 20 December). A series of articles documented how the close relationship among Miami's builders/developers, elected county officials, Zoning and Building Code Board members, and building inspectors over the years was accompanied by a steady reduction in building codes, the acceptance of new building materials and technologies inappropriate for South Florida, and laxity in inspections and enforcement. While researchers have long recognized the critical effects of material culture – in the form of the infrastructure and built environment – on disaster damage, Hurricane Andrew reminded us that to understand these mitigation failures we must appreciate the socio-political ecology that led to their production (cf. Scanlon 1988; Tierney 1989; Peacock 1996).

Heterogeneity and complexity

The complexity of a community's network has consequences for the level and nature of disaster response. No community is able to mobilize sufficient internal resources to respond to a major event, but smaller, less developed areas will be especially in need of supra-local resources. To meet emergency and relief needs, governmental and non-governmental agencies typically flock into the area, bringing workers and supplies. Rebuilding usually requires extensive resources, including not just labor, but new contractors, building suppliers, wholesalers, inspectors, and financial institutions (Barton 1970; Dynes 1974; Wenger 1978; Peacock and Bates 1982). Communities often undergo, at least in the short term, modifications in their scale, complexity, and heterogeneity (Mileti, Haas and Gillespie 1977; Hoover and Bates 1985; Morrow 1992).

A number of past studies have examined community impact, including aggregate change in the number and types of businesses (cf. Friesema *et al.* 1979; Wright *et al.* 1979). However, a host of issues have yet to be fully examined, including a comparison of survival rates and strategies of local, national, and even international businesses, alterations in the nature of employment, and modifications in the size and scale of certain economic sectors. Also, changes in the characteristics and implications of community divisions of labor require further examination. For example, the introduction of new organizations and linkages between a community and its social environment introduces new areas of potential conflict and competition (Faupel 1985 and 1987) and may alter a community's power structure. We report on a short-term analysis undertaken by Dash *et al.* (Chapter 11).

An analysis of the effects of disasters on ethnic populations and businesses represents another fruitful line of research. Disasters do not impact sub-populations equally, yet we have only recently begun to examine disaster-related population shifts (Morrow-Jones and Morrow-Jones 1991). While early research focused on differential mortality rates of ethnic/racial groups, the overall conse-quences for population composition were not examined (Bates *et al.* 1963). Greater population mobility, more dense inhabitation of vulnerable areas, as well as changing ethnic composition, increase the likelihood that disasters will result in ethnic/racial demographic shifts. For example, the implications of increased ability to flee a disaster area as a household recovery option is virtually unstudied. In this book, Girard and Peacock examine ethnic/racial differentials in population movement after Hurricane Andrew (Chapter 10).

Recent Census figures reveal major changes in the heterogeneity of American living arrangements. Households are increasingly composed of non-family residents, single persons, or single parents. The proportion consisting of a married couple (with or without children) decreased from 75 per cent in 1960 to 55 per cent in 1991 (Ahlburg and DeVita 1992). Single adults living alone now account for 25 per cent of all households. About 12 per cent of African-American and 6 per cent of Hispanic children currently live in their grandparents' homes, often with one or both parents. Increasing numbers of the elderly, especially women, live independently. At least 10 per cent of American mothers are raising their children alone (O'Hare 1992) and they are often minority mothers whose economic status is marginal. For example, about one-third of all African-American families consist of a mother and her children with an average annual income less than one-third that of dual-earner families (O'Hare 1992). Not only has household structure changed, but family roles reflect the changing responsibilities of women. The participation of wives and mothers in the workplace has dramatically increased, making the dual-worker family dominant (Ahlburg and DeVita 1992). While this has been accompanied by a slight increase in the domestic involvement of men, women continue to be responsible for most domestic and caregiving tasks (Goldscheider and Waite 1991), often causing serious role overload.

These changing dynamics of ethnicity, household structure, and gender roles impact on the strategies and resources of households in pre- and post-disaster response. Policies need to take into account cultural and structural differences in household resources and patterns of decision making, including the expanded role of women. To guide policymakers, more disaster research should focus on populations reflecting these national trends, particularly since this diversity is especially evident in regions, such as California and Florida, with a high probability of experiencing major natural disasters. Our focus on population diversity is reflected in connection with a variety of topics, including evacuation, emergency housing, women's experiences, family response, and population movements.

The social ecology of communities is modified after a disaster as new groups emerge or move into the area and new linkages among groups are established. In recent decades the literature on emergent social groups and organizations has produced a number of important insights into issues related to future mitigation and recovery (Turner and Killian 1972; Forest 1978; Neal 1984). The ecological network's complexity is altered, particularly among its sub-networks of voluntary and service organizations. Their survival rates and long-term roles in political action and resource mobilization represent important areas for future research. Many of these new organizations and sub-networks, such as the Unmet Needs Committee in South Dade, play important roles in coordinating activities within the community's ecological network.

Coordination

When examining coordination issues, the network interactions of all forms of organizational coordinating structures – not just governmental ones – must be considered. The relations among organizations playing critical roles in disaster preparedness and emergency response are only now beginning to be studied. For example, shifts in interactions as organizations move from an emergency mode to long-term recovery activities call for new roles and demand a new set of sub-networks to oversee coordination; researchers have begun to describe and analyze these shifting network structures (Gillespie *et al.* 1992; Gillespie and Colignon 1993). Averch and Dluhy discuss the particular crises that occurred in Dade County as activities shifted from warning and evacuation to emergency/ restoration (Chapter 5). While addressing leadership problems and key actors involved in this transition, underlying their discussion are the structural issues of inter-organization coordination in times of crisis.

In the United States a host of new private, non-profit organizations have been created with the blessing and even at the behest of government to coordinate rebuilding and recovery efforts. Examples are organizations such as Rebuild LA, established following the 1991 Los Angeles riots, and Florida's We Will Rebuild (WWR), which was set into motion immediately following Hurricane Andrew. These organizations coordinated recovery efforts, not simply by pooling and channeling private funds and public monies to recovery and reconstruction activities, but also because their membership was drawn from the highest echelons of local and national businesses. As such, they wielded extraordinary direct and indirect influence through funding rebuilding efforts, shaping policy and coordinating the activities of key businesses, financial institutions, and government.

WWR was formed following Hurricane Andrew at the request of President Bush. It was headed, administered, and staffed by some of Miami's most influential business leaders, resulting in an under-representation of women and minorities. Soon after its formation, WWR came under fire for its perceived over-focus on long-term business and economic concerns to the neglect of immediate civic needs, including restoring community services to South Dade's families. Protests were heard from various minority groups in the community, including a diverse coalition of community women's organizations which emerged to draw attention to neglected issues of women and families, such as the immediate need for childcare and recreational facilities. While it appears that these umbrella coordinating groups are becoming the national model in communities faced with major disasters, many questions remain unanswered. What is the appropriate role for these powerful organizations? Who do they represent? Should they have control over public, as well as private, funds? What has been their effectiveness in stimulating recovery? Which segments of the community have benefited the most? Issues related to interlocking directorates, the channeling and expenditures of funds, and how membership shapes perceptions of what is determined to be a community need should be explored. Some of these issues are briefly addressed by Enarson and Morrow (Chapter 7), by Dash *et al.* (Chapter 11), and in the final chapter by Morrow and Peacock.

Government agencies are expected to play critical roles in coordinating disaster response. Yet, as discussed earlier, local, national, and regional governments are, to varying degrees, in competition and conflict. Hence, mechanisms for intergovernmental coordination are critical. We address some of the difficulties faced by a small local government, Florida City, in competing for post-disaster resources (Chapter 11). The unique leadership roles needed to support intergovernmental coordination during the emergency response are a major focus of the work of Averch and Dluhy (Chapter 5).

An additional issue concerns the coordination between the government and non-profit organizations. Government agencies tend to be oriented toward dealing with the for-profit sector and, as a result, may fail to facilitate the licensing, regulatory mechanisms, and support for the voluntary groups which have become such an important part of community rebuilding efforts. In Dade County, a host of these organizations, most with religious affiliation, launched extensive programs to help poor households repair and rebuild. Their work was severely impacted, particularly in the early months after Andrew, by the failure of local and state government to recognize their importance and develop policies to facilitate their work. These voluntary groups typically lack political influence, and at each new disaster site are faced with public agencies unfamiliar with their needs. Research is needed on how the work of non-profit organizations active after disasters can be better coordinated and facilitated.

In the United States, coordination within community networks is largely a function of the market, especially as it relates to housing and insurance – two critically important issues in disaster mitigation and recovery. Much of the work which focuses upon the problems with insurance related to housing would easily justify an indictment of the market system as a mechanism for recovery (cf. Squires and Velez 1987). Additional justification stems from the fact that extensive levels of damage can often be traced to inadequacies in the construction materials and techniques used by profit-oriented businesses. However, when comparing impacted communities in the United States with those in centrally planned formal economies, such as the former USSR, our condemnation must be tempered by their failure to provide adequate housing, mitigate against disaster agents, and self-regulate with respect to environmental impact and civil rights policies. Throughout our work, however, we discuss policy-related changes that address more adequate market response to disaster issues.

Competition and its mitigation

In a "free market" system, competition is often lauded as being the driving force behind continued economic development and innovation. However, it is actually *constrained competition*, in which open and violent conflict is kept in check, that is important. In the crisis situation following major natural disasters, unconstrained competition can have a dramatic negative impact on community recovery. In the aftermath of Hurricane Andrew, thousands of poor households lacked sufficient resources to procure for-profit contractors, builders, or service providers. Recognizing the desperate situation, a host of religious and secular non-profit organizations sought to provide reconstruction and repair services. However, as mentioned above, their efforts were initially thwarted by Dade County regulations, which made it very difficult for voluntary construction workers and contractors to work in the area. For-profit contractors and developers saw non-profit groups as competitors and exerted political pressure to block regulatory reforms intended to facilitate the licensing of temporary voluntary labor. As a result, precious time elapsed before sufficient public pressure forced new legislation to support the voluntary agencies. A clearer understanding of these issues, which occur over and over again when non-profits enter new communities, is needed and policies developed to address the problem.

Competition among local non-profit and national voluntary organizations can also be a major impediment to the effective use of limited human and economic resources. The development of formal and informal coordinating groups can lessen conflict, negotiate competition, and facilitate coordination. Following Hurricane Andrew there were a host of attempts to coordinate and reduce competition among these groups, but only a few survived and proved to be truly effective, including a VOLAG which later evolved into an active VOAD (Voluntary Organizations Active in Disasters), an Unmet Needs Committee, and several interfaith coalitions, such as ICARE. Analysis of their goals, organizational structures, effectiveness, internal dynamics, and external relationships is critical to the development of effective mechanisms to lessen competition among them and to better facilitate the use of their vital human and material resources.

Conflict and competition are also evident among governmental agencies, among for-profit organizations, and among the many groups and population aggregates, such as ethnic groups, existing within communities. This can result in delays, blockages, shortages, and waste of limited resources within the field during critical recovery periods. However, disaster researchers must also examine the consequences of competition during all phases of disaster processes. Work by Dynes (1974), Wenger (1978), and Faupel (1987) suggests fruitful lines of analysis. We will address many of these issues in the final chapter of this book.

AN APPLIED FOCUS

This discussion is not meant to be exhaustive, but to suggest issues that flow from a socio-political ecology approach. Community processes occur within exchange networks of contingent relations among the social systems, organized in divisions of labor, constituting the ecological field. Research generated from this perspective targets the flow of people, information, and resources within the network and the competitive environment which occurs as social units with varying amounts of power seek access. Hence, issues of coordination, competition, inequality, changes in patterns of interaction, and field heterogeneity become important dimensions of investigation.

It is our hope that by explicitly developing the unique contribution which disaster research can bring to bear on this theoretical perspective we will make important contributions toward a more comprehensive ecological framework in the social sciences. More importantly, the work of disaster researchers, whether involved in planning, evacuation, mitigation, or recovery, provides an opportunity to contribute to a reformation of the structural perspectives still implicit in much of sociological theory. This is possible not only because we, like other social scientists, are interested in discussing the problems and failures of social systems and networks, but, just as importantly, because the applied focus of our research demands critical evaluation of the effectiveness of these structures. Through further development of a socio-political ecology perspective, it is our hope to stimulate progress toward what Quarantelli (1987) suggested must be our fundamental interest – making communities, and ultimately society, better able to mitigate against and cope with disasters.

NOTES

1 This rejection is based on advances in systems perspectives and neo-functionalism (Alexander and Colomy 1990), constructivism and self-referential systems theory (von Glaserfeld 1984; Maturana and Varela 1980), and work by Pelanda (1982 and 1989) and Bates (1993) at the Laboratory for Socio-political Ecology (cf. Bates and Pelanda 1994). In particular we have drawn heavily from Bates' (1993) critique of the concept of system and resulting reformulation of ecological field theory (cf. Bates and Peacock 1989a).

3

BEFORE THE STORM
The socio-political ecology of Miami
Guillermo J. Grenier and Betty Hearn Morrow

Miami is not a typical US city. The perception is it's too hot, too wet, too violent, and, the real clincher, *it has too many foreigners.* We are used to hearing negative comments – often from tourists returning to land-locked middle America after being here just long enough to take a cruise and develop an attitude. We who enjoy the excitement of Miami have grown weary of responding to comments such as "I can't wait to get back to the United States," or "Miami is too ____" (you fill it in). They are, after all, correct. Miami is different – always on the move and usually moving rapidly in hard-to-predict directions. Even before Hurricane Andrew, this city was used to the heavy-handed media attention associated with international politics, crime, social unrest, and even epic natural disasters.

Miami's uniqueness is too much for many visitors, even some residents. It is a city of contrasts, contradictions, and extremes. Presidential aspirations have begun and ended here. Not many cities have the distinction of serving as the headquarters for international revolutionary and counter-revolutionary activities. The Contra war in Nicaragua was plotted in restaurants and coffee shops in the area known as Sweetwater and clandestine attacks on Cuba continue to be launched from Miami marinas and airports. It has served as a refuge for the famous and infamous, including Prohibition-era gangsters, serial killers, deposed Latin American dictators, sheikhs, rock stars, artists, writers, athletes, and just about anyone seeking to make a new start. As political conditions in the hemisphere vacillate, Miami continues to be the destination of waves of refugees – from the chaotic flotilla of "Marielitos"[1] to the more recent desperate crossings of Haitian boat people and Cuban "balseros" (rafters).

Miami now has the largest proportion of foreign-born residents of any US metropolitan area. With nearly half the population of Hispanic origin and more than one-fifth Black, many from the Caribbean, multiculturalism *is* Miami.[2] In the foreword to *Miami Now!*, Alejandro Portes argues that Miami is America's most internationalized city: "What makes Miami distinct is not the large number of foreigners, other cities like New York and Los Angeles have many more immigrants. It is rather the rupture of an established cultural outlook and a unified social hierarchy in which every group of newcomers takes its

36

preordained place" (Portes 1992: xiii). Here immigrants are reluctant to buy into the expectation that if they start at the bottom, work hard, and play by the rules, someday their children will enjoy the benefits. Whether Cuban, Nicaraguan, Mexican, or Haitian, Miami's immigrants are apt to want their piece of the pie sooner and, what's more, enough get it to make this more than just "American dream" rhetoric (Pedraza-Bailey 1985). But why is this the case? How does the ethnic mix that is Miami play itself out? What are the sources of influence and money? Issues of political and economic power are relevant if we are to understand the context – the socio-ecological field – within which the people of South Florida experienced a catastrophic hurricane. And a glimpse of Miami's history is a useful place to begin.

AN ECCENTRIC PAST[3]

Miami has never been ordinary. It didn't develop at a strategic point along a well-traveled river or railroad route. Unlike Chicago, Cincinnati, or Pittsburgh, it never attracted industrial capital. Nor did it emerge as a seaport competing with New Orleans or the port cities on the East Coast. From its inception at the turn of the century until World War II, Miami was a frontier city. Newer and less traditional, its culture defied classification as either Southern or Northern. With a largely transient population and a high proportion of first-generation residents, the area has always lacked a consolidated socio-political structure and a substantial core of well-established elites. In most respects, it has welcomed newcomers, particularly if they possessed money, dreams, and panache.

Miami was built on the visions of eccentric speculators. In the 1880s, Henry Flagler, millionaire partner of John D. Rockefeller, moved to Palm Beach for his health and found a new career as a railroad and real estate magnate. A decade later he was persuaded to extend his railroad to Miami by Julia Tuttle, an unconventional widow who had left society life in Cleveland, Ohio to forge a new home for her family. She built a house on the northern bank of the Miami River – a site where ancient indigenous peoples of several eras had lived and which would eventually become the hub of downtown Miami. Her dreams and persistence were a critical factor in Miami's early development, making it the only major American city known to have been conceived by a woman (Allman 1987). According to popular legend, when an unprecedented cold winter in 1895 destroyed most of the crops upstate, she seized the opportunity to illustrate the market potential of the warmer climate of South Florida by sending Flagler an orange blossom. While the real story is a bit more complex, it was Tuttle's tenacity, coupled with an offer of free real estate, which convinced Flagler to extend his railroad southward, providing the necessary transportation link for Miami's emerging tourist and citrus industries.

This opportunistic beginning seems appropriate for Miami. As T.D. Allman in his bestseller, *Miami: City of the Future*, expressed it: "The big chill, the orange blossoms, the magnate with an urge to build railroads, the free land to sweeten the deal. Or, in today's idiom: disaster, public relations, the itch not just to escape, but to create new worlds – everything that propels Miami today was there at the creation" (Allman 1987: 133–34).

SNOWBIRDS FROM THE NORTH

The area experienced dizzying growth between 1910 and 1925 as it went from 5,000 to 146,000 residents. After a causeway was built across Biscayne Bay, the filling-in of mangrove swamp to develop Miami Beach began full throttle. It was soon covered with the palatial mansions of rich Northerners – industrialists, entertainers, and mobsters, living side by side. Inland, the grandiose plan of developer George Merrick resulted in the upscale community of Coral Gables. Commercial farming emerged in remote areas south of Miami in what would become the towns of Homestead and Florida City.

Crisis events have always played an important part in shaping Miami's history and this first economic boom came to an abrupt end in 1926 when a major hurricane devastated the city. The Great Depression followed. One effect of this economic downturn was that Jews, previously barred from many of the hotels and private clubs on Miami Beach, were welcome now that their resources were badly needed. By the mid-1930s there had been a significant influx of Jewish working-class and middle-class migrants from the Northeast. This led to a pro-liferation of distinctive small hotels and apartments, particularly in the southern end of Miami Beach in what 40 years later would become the Art Deco district, now immensely popular with European tourists, fashion photographers, and yuppie roller bladers.

The next development boom began with World War II, picking up steam in the postwar decade, when many service men and women who had been stationed in South Florida during the war returned here to live. The commer-cial air travel industry was launched with the establishment of Pan American and Eastern airlines, and communities such as Miami Springs and Hialeah sprang up near the airport. Retirees of all classes came to escape harsh northern winters and settled throughout the area. While the wealthy continued to build winter homes, affordable transportation brought Miami vacations within the reach of growing numbers of middle class American families, boosting the development of a year-round hospitality industry. Blacks from the rural South and the Caribbean were drawn to jobs in the area, settling in segregated communities in Overtown, Coconut Grove, Princeton, and Goulds. Between 1940 and 1950 the population of Metropolitan Miami nearly doubled, rising to 505,000; by the 1960s it had reached nearly a million. With its transplanted groups from the North and South, as well as the Bahamas, Miami already showed signs of becoming an "ethnic cauldron" (Mohl 1983). The next wave of development would be spurred by migration from Latin America and the Caribbean, usually as a result of the rise and fall of various political regimes.

EXILES FROM THE SOUTH

South Florida has always had a close link with Cuba, including a regular ferry service from Key West and an air service dating back to the 1920s. By 1948 Cuba led all countries in the world in the number of passengers exchanged with the United States (Immigration and Naturalization Service 1948). Two deposed Cuban presidents made Miami their home. A prominent Cuban politician of the 1940s built Miami's first baseball stadium. Fidel Castro spent time in Miami in the 1950s and some of his family live here today. Following the 1959 Cuban Revolution and the subsequent failure of the Bay of Pigs invasion, Cuban exiles began arriving in mass, supported by US policies and assistance programs. Some settled in New York and New Jersey, but most preferred the already Latin atmosphere of Miami. By 1980, 52 per cent of all Cuban Americans lived in Dade County (Pérez 1985) and the number of Cubans in Miami rose dramatically once more in 1980 as a result of the Mariel boatlift.

Plate 3.1 Patriotism and spirit were expressed in a variety of ways in multicultural
South Florida
Source: John Kral/*Miami Herald*

The city's emergence as the "capital of the Caribbean" has reinforced immigration trends. Miami is the most desired migration point from the Caribbean Basin, especially among the elite and middle classes. As the United States abandoned the Contra war (which had largely been based in Miami) in the late 1980s, a flow of working class immigrants began, first from Nicaragua, then from other Central American nations, as well as Colombia, the Dominican

Republic, Puerto Rico, and Cuba. Meanwhile, the frustration of democracy in Haiti increased emigration pressures there, despite US policies designed to discourage Haitian crossings. The agricultural enterprises of south Dade County has become largely dependent upon the labor of Mexicans and Mexican-Americans.

At the same time that immigration was increasing dramatically, the influx of northern Anglos slowed to a trickle, becoming negative after 1970. This demographic change was especially evident among the elderly and Jewish sectors. The domestic tourist industry stagnated, as vacationers seeking sun and sea found they could afford Caribbean destinations and families were drawn to Walt Disney World in Central Florida. At the same time, many long-time residents reacted to the city's growing multiculturalism by leaving – a 24 per cent reduction in Anglos occurred between 1980 and 1990. As shown in Figure 3.1, from a 1950s peak at 85 per cent, the proportion of Anglos has been decreasing relatively since 1960 and absolutely since 1970, and is expected to decline to around 30 per cent by the year 2000, barely more than Blacks. This trend is apparent not just in the inner city, where the change is most dramatic, but throughout the metropolitan area, including the suburbs and elite residential communities.

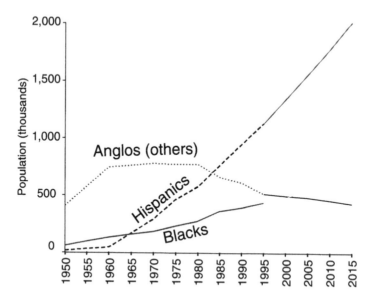

Figure 3.1 Trends in ethnicity in Dade County, Florida
Source: Metro Dade Planning, Development and Regulation Department

GREATER MIAMI TODAY

In general, when people speak of Miami, they really mean Dade County, not the smaller incorporated City of Miami located at its hub. Within the county there are twenty-six incorporated municipalities, including Miami Springs, Hialeah, North Miami, North Miami Beach, Miami Beach, Coral Gables, Homestead, and Florida City. It is important to note that most of the area's population and sprawling territory falls outside of the incorporated communities. The largest governmental body is Metropolitan Dade County. As constituted by "good government" reform in 1957, this strong county government has an appointed manager and, until recently, was administered by a non-partisan commission elected at-large. Within the boundaries of Dade County, each municipality retains some local autonomy; however, in most areas it must abide by standards set by the county. Metro Dade's government is a huge bureaucracy employing more than 35,000 people, excluding the countywide public school system (BEBR 1994).

By 1995, the estimated total population of Dade County exceeded two million and the area had become truly multicultural. Nearly 45 per cent of its residents were born outside the country, about half of whom arrived here since 1980. Cubans accounted for 29 per cent of Dade's population and all persons of Hispanic origin together make up 49 per cent of the total in 1990. About 60 per cent of Hispanics are Cuban, with the others being from throughout the Caribbean and Americas, particularly Nicaragua, Puerto Rico, and Columbia (see Table 1.1, page 10). In comparison, more than 20 per cent of Dade residents report their race as Black, about 20 per cent of which are from the Caribbean (Bureau of the Census 1992).

As South Florida acquired the dimensions of a large metropolis, its economy diversified, becoming an important US commercial port for Caribbean and Latin American markets. Over the last 30 years, Miami has displaced New Orleans as the country's principal trade entrepôt with Latin America. By the early 1980s, 100 multinational corporations had their Latin American head-quarters here and Miami stood second only to New York as an international banking center (Mohl 1985). Currently, 429 multinationals have offices in the Miami area (Beacon Council 1994–95b).

Yet, elements of the old Miami have not disappeared. Anglo men still control the largest firms and dominate many important institutions. The major newspapers, radio and television stations continue to be owned and operated by non-minorities. Interestingly, the "White flight" from Dade County has been class-selective. While Anglo laborers and production workers left, executives and managers tended to remain or move into the area. The elite city of Coral Gables remains principally Anglo, while Jewish influence continues to be strong on Miami Beach. Black Miamians have been involved in a continuous struggle for economic and political power, only in Miami that struggle has taken a particular form. Indeed, in order to understand the socio-political climate at the time of

Hurricane Andrew, we need to consider two significant features – the Cuban enclave and the continued isolation and subjugation of Black Miamians.

THE CUBAN ENCLAVE

Miami's Cuban community is regarded as the foremost example of a true ethnic enclave in the United States. An ethnic enclave is "a distinctive economic formation, characterized by the spatial concentration of immigrants who organize a variety of enterprises to serve their own ethnic market and the general population" (Portes and Bach 1985: 203). The foundation of the enclave is not simply its size or scale, but its highly differentiated nature. In the first place, 42 per cent of all enterprises in Dade County are Hispanic-owned, a percentage second only to Los Angeles, and here they generate far more revenue (Metro Dade 1993). Three-quarters of Hispanic-owned enterprises are controlled by Cubans. The second and most important overall feature, however, is its institutional completeness. In other words, the range of sales and services controlled by Cubans, as well as their penetration into the professions, is so extensive that it is possible for Miami Cubans to completely live and deal within the ethnic community (Portes and Stepick 1993; O'Hare 1987).

The completeness of the enclave insulates the newly arrived somewhat against the usual vicissitudes of the secondary labor market. In contrast to Mexican immigrants, for example, who join the labor market in peripheral sectors of the economy dominated by Anglos and with little informal support, many recent Cuban immigrants enter the labor market largely through businesses owned or operated by earlier arrivals. While wages may not be higher in the enclave, ethnic bonds provide informal networks of support that facilitate learning new skills, access to resources, and the overall process of economic adjustment. These positive implications have resulted in a socioeconomic position which is relatively high in comparison with most immigrant groups (Portes 1987; Portes 1981).

While Cuban entrepreneuralism is impressive, we should point out that most businesses are small and family-owned. Only one out of seven Hispanic businesses in 1990 had paid employees, and together they generated only 30,000 paid jobs (Metro Dade 1993). Latins remain under-represented in the fastest growing industries, especially financial services. Although their representation has increased, they continue to be outnumbered by Anglos in professional and executive occupations.

Cubans have established pivotal political power in Miami, however, exercised through an increasing number of elected officials, as well as influential organizations, such as the Cuban American National Foundation, the Latin Builders Association, the Hispanic Builders Association, and the Latin Chamber of Commerce. By the late 1980s, the cities of Miami, Hialeah, Sweetwater, West Miami, and Hialeah Gardens had Cuban-born mayors. The Miami city manager and Dade County manager were Cuban, as were the majority of Miami commissioners. More than a third of the county's current delegation to the state legislature is Cuban, where their coalition has become influential, and two local Cubans now serve in the US House of Representatives. Nowhere else in the country, nor even in American history, have first-generation immigrants so quickly and so thoroughly appropriated political power.

These economic and political gains have not been made without causing resentment among other groups in Miami, particularly Blacks, who have seen new immigrants come into the community and within a short time achieve the power and success which continues to elude them.

BLACK MARGINALIZATION

From the cavalier way in which ancient Indian burial mounds were destroyed, to the creation of modern Black apartheid, Miami was established on clear principles of White superiority. Allman (1987: 143) expresses it this way: "The new city would have its tragedies; it would have its triumphs. But here was its founding irony; having established this shining city of the future, Miami's creators furnished it with all the darknesses of the past." The economy of South Florida has always been largely dependent upon Black labor. The earliest construction and railroad workers were Bahamian. Black migrants from northern Florida, Alabama, and Georgia formed the core of the agricultural labor force. During the years before the Hispanic influx, most hotel workers were Black. While their labor was needed, their voices were unwanted.

Blacks in Miami were expected to "stay in their place." When the City of Miami was chartered in 1896, Florida law required a petition signed by a minimum of 300 citizens. There were only about 200 White residents at the time, so Flagler took care of this by having his "black artillery" of railroad workers brought to the meeting. After voting for the charter, i.e. in the best interests of Flagler, the workers were sent back over the tracks and their disenfranchisement renewed (Portes and Stepick 1993: 72). City officials were quick to impress upon Blacks, especially the Bahamians who considered themselves social equals, what was considered to be appropriate behavior for "southern Negroes." In the years that followed, racial violence was commonplace (Dunn and Stepick 1992).

The grievances of Miami's Black population are as old as the city itself. In a chapter aptly entitled "Lost in the Fray," Portes and Stepick (1993: 178) document how "The story of Blacks in Miami has always been one of powerlessness, suffering, and frustrated attempts at resistance." Strict Southern segregationism was enforced and its legacy has been hard to erase. At first, most Blacks lived in "Colored Town" (later renamed Overtown) and later in Liberty City, north of downtown Miami, or in walled-off sections of Coconut Grove. As still other Black settlements developed, including Brownsville, Opa Locka, and, in the southern part of the county, Perrine, Princeton, Goulds, and Florida City, segregation was maintained. Before the civil rights legislation of the 1960s, the only Blacks allowed to stay on Miami Beach were live-in servants. Workers in the hotels and restaurants were required to carry identification cards and return to their homes across the bay each night. Entertainment stars appearing on Miami Beach, such as Sammy Davis, Jr., had to stay in hotels in Overtown. Apartheid endured into the early 1960s, when Miami scored about 99 on the residential segregation index (where 100 signifies total segregation), making it the most segregated metropolitan area in the United States (Massey and Denton 1993).

For the most part, the Civil Rights Movement came quietly to South Florida, doing away with the formal institutions of segregation. But just at the point where new opportunities for upward mobility for Miami's Black minority began to occur, the city was transformed by the sudden arrival of Cuban refugees. The rapid economic advance of this group, fueled in part by government programs aimed at national minorities, provided a new focal point for Black discontent – described as the "Miami Syndrome" (Stack and Warren 1992).

The quest by Black Miamians for political strength has confronted two debilitating conditions: weak community leadership and an unresponsive political system. The first was a by-product of typical 1960-era urban renewal programs, especially freeway construction. As in many cities, one positive result of segregation was the emergence of Overtown, a vibrant center of small businesses and professionals serving the local Black population and culture. Urban renewal virtually destroyed Overtown, displacing much of the Black middle class to newly desegregated suburbs, new Black suburban developments such as Richmond Heights, or away from Miami altogether – often to Southern cities, such as Atlanta, offering greater economic opportunity. The net result was an early split between more affluent Black suburbs and an inner city Black underclass.

Socioeconomic class cleavages within the Black community are often apparent and reflected in a perceived lack of support from Black leaders for grassroots racial issues. Similar class differences separate professional Haitians from the masses who support Aristide (Portes and Stepick 1993). Figure 3.2 illustrates the geographical division and separation among Dade's Black communities; none are contiguous. As a result of being spread throughout the county, it is much more difficult to develop common agendas and unified political action. In comparison, Hispanic communities (Figure 3.3) are joined in a wide band extending westward from downtown Miami. Add these together and we have the "Miami Syndrome" (Stack and Warren 1992) – a Black community divided by class, culture, and space which must function under double subordination to Anglos and Cubans. There's little wonder that effective leadership has been slow to develop.

Dade County's metropolitan governance system has traditionally provided little possibility for redressing Black concerns. It is argued that the movement to reform local politics in the late 1950s by developing the nation's first metropolitan government, including at-large elections for all commissioners, suppressed any effective forum for neighborhood and minority differences (Stack and Warren 1992). With some 60 per cent of the area's Black population

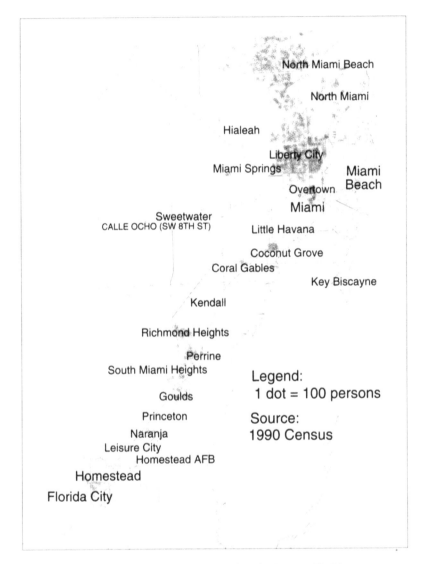

North Miami Beach

North Miami

Hialeah

Liberty City

Miami Springs

Miami Beach

Overtown

Miami

Sweetwater
CALLE OCHO (SW 8TH ST)

Little Havana

Coconut Grove

Coral Gables

Key Biscayne

Kendall

Richmond Heights

Perrine

South Miami Heights

Legend:
1 dot = 100 persons

Goulds

Princeton

Source:
1990 Census

Naranja

Leisure City

Homestead AFB

Homestead

Florida City

Figure 3.2 Black population of Dade County, Florida

residing in unincorporated Dade County, and much of the remainder in the City of Miami with its large Hispanic majority, the chance of generating effective Black political representation has been non-existent.

Four riots during the 1980s crystallized a widespread anger among Black Miamians over both their failure to keep pace economically with other groups and their lack of political voice (Herman 1995). The response of city elites was to create a series of economic and social programs designed to shore up Black

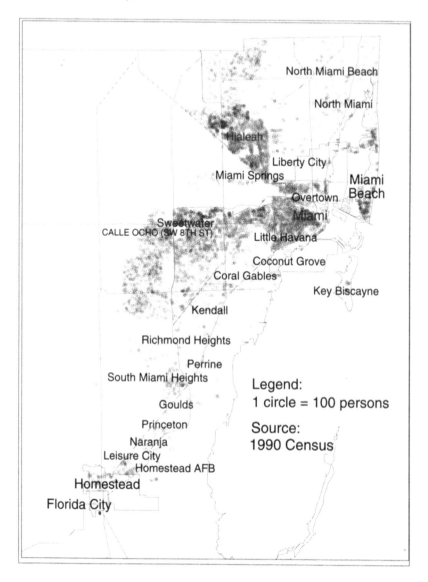

North Miami Beach

North Miami

Hialeah

Liberty City

Miami Springs

Miami Beach

Overtown

Miami

Sweetwater

CALLE OCHO (SW 8TH ST)

Little Havana

Coconut Grove

Coral Gables

Key Biscayne

Kendall

Richmond Heights

Perrine

South Miami Heights

Legend:
1 circle = 100 persons

Goulds

Princeton

Source:
1990 Census

Naranja

Leisure City

Homestead AFB

Homestead

Florida City

Figure 3.3 Hispanic population of Dade County, Florida

neighborhoods (Reveron 1989). The task of rejuvenating the Black community was daunting, but there were some successes. Compared to a decade before, by 1987 the number of Black businesses had more than tripled (Bureau of the Census 1990). Under a set-aside program Black contractors began receiving county work. In a gesture which might well symbolize what it takes to develop successful businesses in Miami's Black communities, in 1990 Otis Pitts received the MacArthur Foundation's "genius" award for building a shopping center on the border of Liberty City and Little Haiti which employed 130 people (Viglucci 1990). In general, development projects have met with limited success and gains in employment have been modest (Dunn and Stepick 1992). In 1991, Black businesses in Dade County numbered about 6,800, but their average sales and receipts ($40,934) was about half that reported for Hispanic businesses (Metro Dade 1993). Black firms accounted for only 0.5 per cent of the total sales and receipts in Dade County – a considerably lower share than is typical of Black-owned business statewide or nationwide.

Gains by Black Miamians have paled by comparison with the economic and political success of Cubans. When Martin Luther King, Jr. visited Miami in 1966, he noted Miami's racial hostility and warned against pitting Cuban refugees against Blacks in competition for jobs (Porter and Dunn 1984). While competition is difficult to document, in the area of government minority contracts, Cubans clearly prevailed over Blacks. Between 1968 and 1980, the Small Business Administration (SBA) cumulatively dispersed 46.6 per cent of its Dade County loans to Hispanics and only 6.0 per cent to Blacks. The situation actually worsened after the riots, when nearly 90 per cent of the SBA loans were awarded to Hispanics or Whites. In 1990, Dade County discontinued its set-aside program.

While Miami's racial and ethnic profile is often portrayed as a tripartite division of Anglos, Hispanics, and Blacks, it is important to note once again the cultural diversity existing within Black Miami. The second major immigrant group in Dade County are Caribbean Blacks. Largely because of Caribbean immigration, Dade's Black population grew by 47 per cent between 1970 and 1980 (a growth rate exceeded only by Atlanta) and another 42 per cent in the next decade (Dunn and Stepick 1992; Bureau of the Census 1992). About 14 per cent of Miami's Black population in 1990 were Haitian.

The first refugees from the repression and poverty of Haiti arrived by boat in 1963. Their request for political asylum was denied, and the second boatload did not come until 10 years later (Stepick 1992). Conditions in Haiti worsened and between 1977 and 1981 it is estimated that as many as 80,000 Haitians came by boat and plane. Those not forcibly returned to Haiti could expect a cold welcome in South Florida. Not officially recognized as refugees, most were ineligible for social services or work permits. Ill-equipped to make it in competitive South Florida, the average Creole-speaking Haitian immigrant has less than six years of formal education and possesses few marketable skills. There has been no well-established enclave to receive them. Most assistance has come from churches and private organizations, such as the Haitian Refugee Center established with help from the National Council of Churches.

In one respect, the Haitian immigration pattern mirrored the Cuban – an early wave of middle and upper class refugees from the Duvalier regime in the 1960s and 1970s has been followed by poorer immigrants. While shops, restaurants, and other small businesses have emerged, Haitians have not been able to emulate the success of the Cuban enclave. The reality of their situation, including US immigration policies to stop them, the lack of official refugee status with its concommitant social and financial assistance, and the social discrimination linked to being Black in America, has resulted in a very different opportunity structure.

Relations between Haitians and Miami's native Blacks have been strained by cultural differences and perceived economic competition. Rather than identifying with the racism Haitians are experiencing, many African-Americans reject them as unwanted competitors in the tight job market. In general, Haitians have tried to avoid being identified with Miami Blacks since they want to rise above what is perceived to be a down-trodden group (Stepick 1992).

In recent years, several developments have started to unite the Black community. A major incident resulted from the treatment Nelson Mandela received when he visited Miami in 1990. Unlike the civic honors bestowed at his other stops, the official Miami reception was decidedly cool – the product of Cuban and Jewish outrage at his refusal to disavow his cordial relations with both Fidel Castro and Yasser Arafat. Most Black elected officials did not take a stand on this issue, an example of the divisiveness sometimes arising within the Black community. A group of Black professionals and community leaders organized a boycott, calling on national organizations with conventions scheduled in Miami to take their business elsewhere. The boycott had an impact. Before the end of 1990, thirteen organizations, including the American Civil Liberties Union and the National Organization of Women, had canceled Miami conferences. In all, Miami is estimated to have lost over $60 million in convention-related business. The agreements which ended the boycott in 1993 included the establishment of a scholarship program for Black students to receive training and subsequent management level employment in the local tourist industry, support for the development of a new Black-owned hotel, and several retroactive statements honoring Mandela.

Recent electoral reform has been the most promising development in furthering the interests of the Black community. A federal court ruling in late 1992 ordered Metro Dade to replace its at-large commission immediately with members elected from individual districts, in order to guarantee minority representation. Following the establishment of thirteen districts, the next election resulted in a new Metropolitan Dade County Commission composed of four African-Americans (including the Chair), six Hispanics and three Anglos – a representative commission after all these years. South Florida currently has two African-Americans serving in the US Congress and six state legislators. Blacks serve on the current city commissions or councils of Miami, El Portal, Opa Locka, and Florida City – the latter two Black communities also have Black mayors (Metro Dade 1994a).

Increased representativeness does not easily translate into political power, as was made abundantly clear during the 1994 choice of a new Metro Dade county manager. To select a replacement for the outgoing Cuban-American, both the

Hispanic and Black communities mobilized forces. Going into the election the five Cuban-American commissioners were lined up squarely behind their Hispanic candidate, while the four African-Americans and one Anglo backed the African-American candidate. The decisive vote for the Hispanic was cast by a politically astute Anglo who had been supported by the Cuban community three years before in her successful bid to become the first woman president of the Florida Senate (Filkins 1994). In a metropolitan area in which ethnic groups – especially Blacks – are spatially segregated, the existence of an institutionally complete community among Cubans makes it highly unlikely that the mutual understanding necessary to work toward a common agenda will be realized.

What does all of this tell us about the countywide context in which ethnic groups experienced Hurricane Andrew? It can generally be assumed that Blacks, as a group, were in a disadvantaged position when the storm struck, compared to either Anglos or Hispanics, at least Cubans. At the household level we would expect them to be more heavily represented among those living in sub-standard housing, especially low-income rentals, which sustained extensive damage. Black families were more likely to be poor, and thus without the personal resources, such as insurance and savings to recover. At both the household and community levels, it was to be expected that they would be operating at an economic and political disadvantage in the highly competitive recovery environment.

While Black communities and their leaders lacked political power going into the storm, we would be remiss in not mentioning the key roles played by several community leaders in Hurricane Andrew recovery efforts. The new political landscape of Dade County has resulted in a dramatic increase in Black office holders at the county, state, and federal levels. As we will discuss later, several key people emerged from that group to spearhead community recovery and redevelopment efforts. Black churches were instrumental in providing relief and recovery services. Over the last three years there have been many instances when Black communities, perceiving they were being left out of planning efforts, challenged community groups, such as We Will Rebuild and the Metro Dade County Commission, to better represent their interests.

WOMEN IN MIAMI

When discussing those likely to be at a disadvantage in the competitive post-hurricane environment, women must be considered as a group lacking power within the Miami community. While Julia Tuttle's vision may have promoted the creation of Miami, its recorded history has little to say about women's contributions. The region does have a few modern-day heroines, such as environmentalist Marjorie Stoneman Douglas, historian Arva Moore Parks, and Black businesswoman and civic leader Athalie Range. For many years women have served (not proportionally) on school boards and local governing bodies. It is only in recent years, however, that they have started to make significant gains in influential political offices and judgeships. There are currently two women in the US Congress, as well as six members of the Florida legislature and four Metro Dade school board members (Dade County Commission on the Status of Women 1995).

About 30 per cent of Dade businesses are owned by women, but they tend to be small operations, both in terms of employees and revenues (Metro Dade 1993). In every comparison, the gender disparity exceeds that for the nation as a whole. Miami ranks sixteenth in the nation in women-owned businesses per capita (ACBJ Research Report 1995).

As commonly occurs following a disaster, women were critical to the survival and recovery of individual households after Hurricane Andrew. They dominated as community service providers, yet were rarely in positions with sufficient power to effectively influence the direction of community response. As we shall discuss in Chapters 7 and 8, any differences they were able to make were typically hard-won.

A RELEVANT SETTING

From this overview of Miami's history and current social climate, it should be clear that this is a complex community. It provides an excellent example in support of our theoretical argument, laid out in the last chapter, that a community should not be thought of as a unified system, but as a network of systems. There is no consolidated, all-powerful socio-political structure in Dade County, although certain groups and constituencies hold sway over specific issues. The environment is one of often fierce competition, interspersed with fragile, ever-changing coalitions. Old power, as represented by Anglo business-men, is being usurped by new factions, typically Cuban, who are not themselves of one mind. Among the previously disenfranchised groups, Blacks and women, there are competing factions even as they attempt to challenge the status quo. At the organizational level, businesses, not-for-profits, public and private agencies, cities, unincorporated neighborhoods, and neighborhood groups compete for power and for shrinking public dollars.

The diversity of South Florida provides a rare opportunity to examine ways in which experiencing a major disaster, from preparation through recovery, is affected by the relative position of the social actor – be it an individual, house-hold, or community – within the larger social and political mosaic. It should be expected that attributes of race, ethnicity, class, and gender, as well as circumstances and location within the larger field, influence the outcomes. The fact that most of the heavily impacted areas fell within the unincorporated areas of Dade County allowed us to look at the possible effects of the position of a neighborhood or local community *vis-à-vis* the larger political context. And, of course, inter-governmental relationships after a disaster are always interesting, including the interactions among cities, between cities and Metro Dade, between Metro Dade and the State of Florida, as well as between the state and the federal government. In short, the unusual milieu that is Miami provides an excellent setting from which to argue for an ecological approach to the study of disasters.

NOTES

1 In 1980, in order to relieve internal pressure, Fidel Castro opened the port of Mariel in northern Cuba for anyone who wanted to leave the country. This unleashed a flotilla of boats as Cubans living in South Florida went to pick up relatives and others. The result was a mass migration of over 150,000 Cubans, who became known in South Florida as Marielitos.
2 For a more complete analysis of Miami as a multicultural city, see: Grenier, Guillermo, and Stepick (1992); Portes, Alejandro, and Stepick (1993); and Allman (1987).
3 Many excellent books have been written about Miami's rich history. See, for example: Muir (1990); Parks (1991); and Peters (1984).

4

WARNING AND EVACUATION
A night for hard houses
Hugh Gladwin and Walter Gillis Peacock

An interview with a 14-year-old boy residing at the Campbell Drive tent city whose family had evacuated from their mobile home to a shelter:

> At first we thought about Southridge, but when we heard they were building something in front of it, we decided either Richmond Middle School or Miami-Dade [Community College]... My mother had passed by Richmond earlier and it was packed already. And then we decided to go to Miami-Dade.

At what point Sunday did you decide [to evacuate]?

> At first we were all deciding that we were going to the shelter. Then we found out we weren't in an evacuation zone, but me and a couple my friends were saying "I'd rather be safe than sorry", so we should go. My grandmother wanted to stay because she knew that it wouldn't be comfortable being with a lot of people in Miami-Dade. But if we would have stayed we would have most likely been dead because our house was damaged, totally... We thought that something bad was going to happen to our house.

What made you feel that way?

> When we saw that the hurricane built to a category 5 [actually it was on the boundary of becoming Category 5]. In a category 4 we would have probably stayed. We found out it was a category 5 and then later around 2 p.m. Sunday, we were deciding to go, but my grandmother still wanted to stay... my grandmother, my grandfather, me, and my mother at home and everyone else to go to Miami-Dade. I didn't want to do it, so I was upset when I heard that. And then we decided, all of us, let's just go... We left around 7 p.m. And the hurricane hit at 5 a.m.

[Younger cousin of first informant, about 6 years old:]

> We should tell you about what happened inside. I fell asleep inside Miami-Dade. And then, when I heard like crashing everywhere I woke up and I said, "Is it here?" and my mother said, "It's here". And I got, and I got scared. Then we heard something falling down – the roof was falling down. And I was shaking a lot and they had to pass us [outside the auditorium] along

those ropes. [The auditorium had to be evacuated during the storm when its roof began to break up.]

When asked what they most wanted, these children who had seen what was left of their homes after Andrew ripped them apart all agreed that what they wanted now was to be able to live in, as one put it, "hard houses."

HURRICANE ADVISORIES

The following excerpts are taken from the texts of two National Hurricane Center (NHC) advisories, issued 22 and 23 August 1992. The first, on Saturday, reports an increase in Hurricane Andrew from a Category 1 to 2 hurricane. The second, only 24 hours later, reports the storm on the boundary between Category 4 and 5. About 15 hours later, much of south Dade County had been virtually destroyed.

Bulletin – Hurricane Andrew Intermediate Advisory Number 23a
National Weather Service Miami Fl 2 p.m. EDT Sat 22 Aug 1992
Andrew continues moving westward . . . A hurricane watch continues for the Northwest Bahamas . . . Andrew continues to strengthen and is moving toward the west near 14 mph. Maximum sustained winds are now near 95 mph. And some further strengthening is likely during the next 24 hours. The westward movement is expected to continue through Sunday increasing the threat to south and central Florida. Interests in that area should closely monitor future advisories on this hurricane.

Hurricane Andrew Special Advisory Number 29
National Weather Service Miami Fl 2 p.m. EDT Sun 23 Aug 1992
Hurricane warnings remain in effect for the central and northwest Bahamas and the Florida east coast from Vero Beach southward through the Florida Keys to the Dry Tortugas . . . All precautions to protect life and property including evacuations ordered by emergency management officials should be rushed to completion. At 2 p.m. EDT the center of Andrew was . . . about 280 miles east of Miami, Florida. Andrew is moving toward the west near 16 mph and this motion is expected to continue for the next 24 hours. Hurricane conditions will spread over the northwest and central Bahamas today and on the present course should reach the southeast coast of Florida in the predawn hours on Monday. This is a dangerous Category 4 hurricane with maximum sustained winds near 150 mph. Some fluctuations in strength are likely before landfall. Hurricane force winds extend outward up to 30 miles from the center and tropical storm force winds extend outward up to 105 miles. Storm surges of 7 to 10 feet above normal tides are possible for the Florida east coast and Keys near to where the center makes landfall in South Florida with possible heights of 9 to 13 feet in Biscayne Bay.

Facing the destructive and life-threatening potential of a strong hurricane can certainly create a common sense of identity. Most Dade County residents, just prior to Andrew, were involved in similar urban hunting and gathering activities – seeking out food, water, gas, and other necessities. We became members of the same symbolic community: a community of fate, a community whose fate was in question.

This image of Miami as a common community unified by a single threat was furthered by the media. Television weather forecasters, such as Bryan Norcross, did not wait for official hurricane warnings; by Saturday morning they were urging people to consider seriously the magnitude of this storm. While the area had been threatened many times before and spared, this time our luck might have run out. Complacency was not a good option with a storm of this intensity bearing down. Non-stop television coverage kept Dade's households abreast of the storm's progress, providing information on what to expect, concrete directions for actions, when evacuation decisions should be made and what their options were. It showed people hard at work preparing for the storm. Overall, the presented picture was one of a community displaying systematic and singular purpose.

This image of an integrated community acting in a coherent, purposive manner is one held by many officials. As Averch and Dluhy argue in the next chapter, those responsible for "managing" disaster response often operate as if the community were a single bureaucratic "system." This image guides their "rational decisions." And, in theory, it works in the sense that substantial proportions of a population often heed official warnings and watches.

In reality, however, cohesive compliance is rarely the case. There is slippage, error, and uncertainty. Except under extreme circumstances, households cannot be compelled to evacuate or to remain where they are, much less to prepare themselves for the threat. Even under extraordinary conditions many households have to be individually located and assisted or forced to comply. Segments of a population may fail to receive, ignore, or discount official requests and orders. Still others may not have the resources or wherewithal to comply. Much will depend upon the source of the information, the consistency of the message received from multiple sources, the nature of the information conveyed, as well as the household's ability to perceive the danger, make decisions, and act accordingly. Disputes, competition, and the lack of coordination among local, state, and federal governmental agencies and between these agencies and privately controlled media and other businesses can add confusion. Business and governmental agencies that refuse to release their employees and suspend normal activities can add still further to the confusion and non-compliance.

An urban area consists not of an integrated system, but rather an ecological network containing many constituencies which can respond in different, seemingly unpredictable, ways to impending crisis. Even the existence of a centralized emergency management bureaucracy cannot insure that individual households will receive accurate information, properly assess the danger, formulate decisions, marshal the necessary resources, and respond as requested. This chapter focuses on how hundreds of thousands of households in Dade County – home to nearly two million residents – reacted to the relatively rapidly developing threat posed by Hurricane Andrew. The issues surrounding three topics will be examined: household preparation, evacuation, and the consequences of this experience for future disaster response. First, we focus on the primary sources of household information and subsequent household reactions. We will seek answers to the following questions: how and when did they learn that Andrew was a dangerous storm likely to strike South Florida; at what point did they start to take actions to protect themselves and their property; what types of activities did they undertake; and what are their overall assessments of their preparation and the media's job?

The second issue to be addressed is evacuation. For people living in low-lying coastal areas evacuation is critical for survival. Yet, from our data we know that approximately 27 per cent of the households along the beach did not evacuate. If Andrew had directly struck coastal barrier islands like Miami Beach, the casualties from storm surge and wind destruction could have been very high. Afterward, the August heat and humidity would have exacted a toll for survivors, particularly the elderly, trapped without air-conditioning in buildings with non-functioning elevators. Sand drifts blocking ground-level doors would have hampered rescue activity and further endangered life. Fortunately, for the almost half-million people under evacuation orders, this grim scenario did not take place. Andrew did not hit the dozens of glass-surfaced high rises along the coasts of Miami Beach and Fort Lauderdale. Instead, it went south where population and business concentrations were much lighter. Nevertheless, the threat forced most residents to deal with the dilemma: should we stay or should we leave? This chapter will seek to understand the factors shaping this household decision.

The final issue concerns the consequences of Andrew for future household decisions. This experience has, at least in the short term, shaped the future evacuation and preparation behavior of South Floridians. In general, we might anticipate that Andrew has made them more likely to prepare their homes, but perhaps less likely to evacuate in the face of future storms. Part of this may stem from the relative ambiguity of the statement, "we have experienced a hurricane." Many households were forced to make an evacuation decision and were certainly part of the "community of fate" created by Andrew. However, a relatively small proportion of those actually under evacuation orders fully experienced its force. We end this chapter with an examination of the potential consequences of Andrew on future household evacuation decisions.

THE MEDIA AND OTHER SOURCES OF INFORMATION[1]

There is considerable research literature on the process people go through in receiving warnings of possible impending disaster, evaluating the threat, and deciding what to do (Drabek 1986: 70–132). Evaluating a potential threat is a complex, socially mediated process (Drabek 1969; Quarantelli 1980; Mileti 1975; Perry, Lindell and Greene 1981; Perry 1985). Individuals in households and other social groups bring their differing experiences to bear in discussing a threat and evaluating its likelihood and potential harm. This process frequently entails a considerable inertia as individuals express disbelief and have to be convinced through discussions with others (Drabek 1986: 101–2). Pronounced inertia might have been expected in the case of Andrew. Decades had passed since South Florida had experienced a major hurricane and in the interim there had been only limited hurricane preparation activities aimed at familiarizing the general public with emergency management and evacuation plans.

In the case of Hurricane Andrew, the official predictions were necessarily tentative. Due to the restrictions built into warning policies, as well as the storm's rapid development, the time between its formation as a hurricane and subsequent landfall was relatively short. The winds reached 75 mph, making it an official hurricane, early Saturday morning; by noon the NHC had declared a Hurricane Watch for South Florida; at 5 p.m. it estimated a 21 per cent strike probability for Miami. Hurricane warnings, which officially set evacuation plans in motion, are by policy not issued until 24 hours before an estimated strike. However, officials in Dade and Monroe counties decided to initiate evacuation orders Saturday to take effect Sunday morning. Despite the relatively short lead-time, our data show that over 80 per cent of Dade's residents had started preparations by Sunday morning. Over 95 per cent felt they had received the information necessary to prepare. How do we explain this high level of preparation response within such a short time-frame?

One factor that must be kept in mind throughout our discussion is that the preparation period occurred over a weekend when individual routines are not as encumbered by work and other major responsibilities. Weekend schedules may have allowed for greater response rates. Further, the extreme speed of the storm's development may have alarmed people, pushing for action. Yet another major factor is that the media, particularly television, was critical in transmitting the warning of the storm.

Generally speaking, the public does not have the ability to track a storm, nor do they have a direct relationship with national weather service personnel. The media becomes the key source for this information. As Perry and Greene say,

> where community involvement of emergency officials is low, and channels of official to citizen communication are limited primarily to mass media, the media take on a part of the official function and evoke similarly high levels of warning belief among citizens.
>
> (Perry and Greene 1983: 52)

As a source, television in particular has two characteristics which facilitate its messages being taken seriously. There are only a few television channels, and since their weather persons get their information primarily from the National

Hurricane Center, there is little contradiction. Indéed, NHC personnel, such as Dr. Bob Sheets, often directly delivered the information on all the major networks. And since a great number of people kept their television sets tuned to the continuous hurricane coverage all weekend, the message was repeated incessantly. All these factors are important because the "more accurate and consistent the content across several messages, the greater the belief . . . Belief in eventual impact increases as the number of warnings received increases" (Mileti 1975: 21).

The significance of the media was evident in respondents' answers to our question about the relative importance of various sources for hurricane preparation information. More than one source could be chosen, but almost 85 per cent of households gave television as their most important source of preparation information. Perhaps not surprisingly, household members with past hurricane experience ranked second (50.9 per cent), followed by the radio (49.9 per cent), relatives not living in their household (34.7 per cent), friends (31.3 per cent), and the newspaper (23.5 per cent).

The central role of television cannot be overemphasized. Most people were based in their homes over the weekend with their television sets turned on. The non-stop hurricane coverage created in South Florida a *virtual community* as compelling as events like the assassination of President Kennedy, the first moon landing, and the Watergate hearings. A big difference between these events and waiting for Andrew, however, was that everyone watching was an actor. The messages had immediate relevancy to their lives.

In a community with sizable proportions speaking English as a second language, if at all, non-English media sources were also important. Over 14 per cent of our respondents who relied on television depended exclusively on channels broadcasting solely in Spanish. Even greater language diversity occurs on radio. Although its full coverage began later than television, for the nearly half of our respondents who indicated radio was also important, 65 per cent listened to English-language stations and 32 per cent to Spanish. The remainder listened to both, or to another language such as Haitian Creole.

When asked to rate the overall job of the media in providing information to help prepare, 67 per cent rated it as excellent, 28 per cent said it was good, and only 5 per cent rated it as fair or poor. For the vast majority of our respondents, as well as 97 per cent of those who felt they had received all of the information they needed, there was strong sentiment that the media had done a good job of providing important and relevant hurricane information before the storm.

WHEN PREPARATION STARTED

As mentioned before, the advent to Hurricane Andrew's strike fell on a weekend, which may have insured earlier preparation time. By Friday evening approximately 22 per cent of surveyed households had begun preparation and by Saturday evening that percentage had more than doubled to almost 59 per cent. Sunday morning, the timing of the official Hurricane Warning, was marked by the largest increase in households beginning preparation (25 per cent), bringing the total to almost 84 per cent of all households.

One way of comparing preparation differences among household types is by examining the mean interval between the beginning of preparation and the hurricane's arrival. As shown in Table 4.1, single elder households started the earliest, nearly 41 hours before the storm, followed closely by households solely inhabited by elderly couples. These early preparers likely had few, if any, employment obligations, allowing them to begin somewhat sooner. Bunched together in the range of 31 to 35 hours are: households composed of two or more unrelated people; adult couples; adults and children; single-female parent households; households composed of a single adult or related non-couple adults; and two or more non-couple relatives living together, such as a younger adult living with a parent. The statistical significance of either being an elderly household or having an elder member in the household is evident.

Plate 4.1 People waited up to four hours in line at grocery stores getting ready for Andrew's arrival
Source: Bill Frakes/*Miami Herald*

Household differences with respect to race/ethnicity, income, living in a single-family dwelling, or living in an evacuation zone had little relationship to preparation time. Homeowners tended to start preparation slightly earlier, but not by enough to be statistically significant. This lack of difference may indicate that personal safety rather than the desire to protect property was the main motivator. We also looked at different media sources to see if certain television stations, particularly those providing more specific instructions and guidance, had an effect on preparation time. We found that households relying on Channel 4 (meteorologist Bryan Norcross), Channel 10 (meteorologist Don Noe), or

Table 4.1 When households began preparing for Hurricane Andrew

	Mean interval between start of preparation and storm's arrival (hours)	N	F or t Ratio
Household type			
Single elder	40.9	52	2.83[a]
Couple elder	38.5	64	
Non-related household	34.7	59	
Adult couple	33.3	188	
Adults and children	32.8	462	
Single mother with children	31.8	149	
Single adult	31.7	92	
Related non-couple adults	31.4	105	
One person over 65 years?			
Yes	39.5	116	4.12[a]
No	32.6	1055	
Single family dwelling?			
Yes	33.2	721	0.10
No	33.3	520	
Homeownership			
Own	33.8	816	1.46
Rent	32.3	420	
Race/ethnicity			
Anglo	34.5	427	1.21
Black	32.9	178	
Hispanic	32.7	591	
Household income			
Under $20,000	34.1	367	0.86
$20,000–$50,000	32.6	450	
Over $50,000	33.8	254	
Live in evacuation zone?			
No	33.2	830	0.39
Yes	32.8	352	
Hurricane experience			
Yes	33.9	800	1.78[c]
No	32.0	436	
TV channel most relied on?			
Channel 4, 10, or 23	34.9	617	2.55[b]
Other	32.3	537	

Source: FIU Hurricane Andrew Survey
Notes: a p≤.01
   b p≤.05
   c p≤.10

Spanish-language Channel 23 tended to begin preparation about three hours earlier than those watching other channels. Perhaps respondents who did not cite one of these stations had a less focused source of hurricane information.

Households with members who had previously experienced a hurricane started nearly two hours earlier; however, this was not statistically significant.

The positive effect of prior hurricane experience is mentioned frequently in the literature on hurricane preparation, predisposing people to start earlier (for example, Baker 1991: 302–6). This is, of course, also correlated with age in that older people are more likely to have more hurricane experience, especially in the case of Andrew when about 30 years had passed since the last major storm. Age also predisposes people to be more conservative, to start preparation earlier, and to want to avoid the disruptions of evacuation. Indeed, in more complex analyses to determine the relative influence of the various factors from Table 4.1, age and relying on the previously mentioned television stations were more important than any other factor, including previous hurricane experience. However, these findings must be interpreted within the context of the very high levels of early preparation across all households and the specific types of preparation actions being considered. With overall high numbers, there were very few statistically significant differences among different types of households in terms of when they started preparing. We can speculate that had Hurricane Andrew approached during the week, employment and school situations would have impeded preparation activities and there may well have been greater variation among households.

PREPARATION ACTIVITIES

Households undertook a variety of activities to prepare for Andrew. Our survey results indicate that 97 per cent had candles, a working flashlight, or a gas-powered lantern on hand; 95 per cent bought or prepared drinking water; 90 per cent had a working battery-powered radio; 89 per cent brought inside or secured loose objects; 87 per cent bought canned or non-perishable food; 62 per cent had charcoal or a gas-powered stove; and 56 per cent protected their windows using shutters or boards. As might be expected when the rates are so high, there was limited variation among different types of households and most of the variation was related to window protection. On a more open-ended question, respondents were asked whether there was anything they had wanted to do, but were unable to because they lacked the time, money, or supplies. The most frequently mentioned answer (17 per cent) was boarding or shuttering windows. Furthermore, when asked if there was anything they would do differently next time, the two most frequent answers were to evacuate (20 per cent) and to shutter their windows (14 per cent).

In light of these findings it can be concluded that, when pressed by the media to consider the seriousness of the approaching storm, people in South Florida did a strenuous and fairly complete job of preparation. However, it is also important to note that most of these activities are relatively easy to accomplish and have more to do with anticipating the lack of services following impact than with mitigating against damage. Indeed, the proportion of households who shuttered their windows – the one activity which could make a significant difference in reducing the loss of life and property – was relatively low. We

60

speculate that it would be even lower had more specific questions been asked regarding how many windows were covered and how they were covered.

The costs to life and property of a major hurricane can be reduced with proper building construction. Effective mitigation involves instituting and enforcing building codes requiring hurricane-resistant construction, retrofitting existing housing, and having appropriate window coverings ready to install. There is rarely enough time to buy materials and make adequate installations during the hectic period when a major storm is making its final approach. These substantial measures demand time, money, planning, incentives, and legislation. If FEMA, the state, and insurance companies are serious with their current rhetoric to shift the high costs of disaster recovery to the up-front costs of paying for mitigation, then more initiatives are needed to help households and communities prepare for South Florida's inevitable future hurricanes.

THE DECISION TO EVACUATE

In Dade County, evacuation areas expand depending on the strength of a hurricane. In any hurricane threat, however, evacuation is required for coastal islands, most areas immediately along the coast, and mobile homes. The most extensive evacuation area begins with a Category 3 storm (111–130 mph wind or higher) and adds a rather wide swath near the coast and much of the southern portions of the county, which are very low and subject to flooding. In the case of Andrew, since official evacuation orders went into effect Sunday morning, the

Plate 4.2 Evacuation of Miami Beach before Hurricane Andrew
Source: Rick McCawley/*Miami Herald*

evacuation areas were being finalized in the minds of many at about the same time as the orders were to be acted upon. By 11 p.m. Saturday evening Andrew had strengthened to Category 3, although this news may not have reached some until they woke up Sunday morning. By 8 a.m. advisories reported winds at 120 mph – Andrew was going to require evacuation over the maximum area.

For households residing along the shoreline or islands, as well as those in mobile homes, the situation should have been more or less evident since Saturday afternoon. On Sunday morning many households further inland were still trying to figure out if they needed to evacuate; nevertheless, by early Sunday afternoon it was clear that Andrew was a very serious storm. Thus, whether a household was located on or off the shore, it is very likely that the way most people were thinking and talking about evacuation changed by Sunday afternoon. The approaching hurricane suddenly became the "Big One." People all over South Florida, not just those near the coast, were now worrying. Although the timing was very late for trying to move far, according to some reports, more than a half million people left their homes seeking safer locations. Of these, it is estimated that 80,000 went to shelters (*Miami Herald* 1992a). Many others went to the homes of family or friends, to hotels, or left the area.

Our survey data on household evacuation are presented in Table 4.2 according to each respondent's self-report of evacuation zone status, as well as whether the household's ZIP Code placed it in a coastal evacuation zone.[2]

Table 4.2 Evacuation comparisons from Hurricane Andrew

	Sample size	% evacuated
Household in evacuation zone:[a]	369	(29.8)
Entire household evacuated		53.8
Some evacuated/some stayed		4.5
Entire household stayed		41.7
Not in evacuation zone:	871	(70.2)
Entire household evacuated		11.7
Some evacuated/some stayed		3.9
Entire household stayed		84.4
Household in coastal evacuation zone:[b]	84	(6.4)
Entire household evacuated		70.8
Some evacuated/some stayed		2.0
Entire household stayed		27.2
Not in coastal evacuation zone:	1216	(93.6)
Entire household evacuated		21.1
Some evacuated/some stayed		4.1
Entire household stayed		74.8

Source: FIU Hurricane Andrew Survey
Notes: a Self-report
        b Based on ZIP Code

Coastal evacuation zones are the high-risk areas most susceptible to damage, particularly flooding, regardless of a storm's strength. These data have also been mapped by ZIP Code area in Figure 4.1.

The number of responses within each ZIP Code is small, showing considerable sampling fluctuation; however, the data allow a rather precise look at evacuation patterns. The areas are shaded according to the per cent of house-

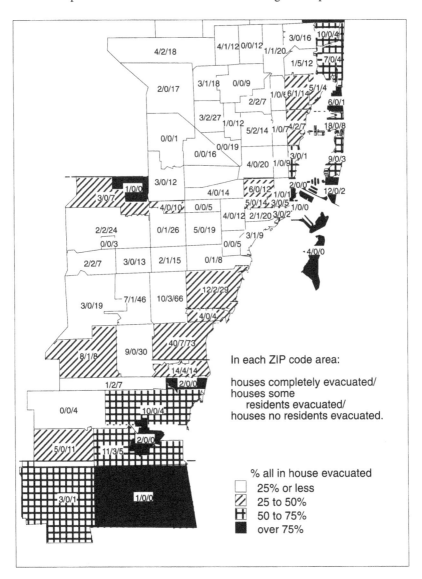

Figure 4.1 Hurricane Andrew evacuation: Dade County households interviewed by ZIP Code

63

holds in which the entire household evacuated. Numberless areas are where random digit dialing did not produce a sample household. The three numbers in each ZIP Code area indicate: the number of households where everyone evacuated, the number where some members evacuated, and the number where no one evacuated. While by themselves the individual numbers may fluctuate, an overall pattern emerges.

Compliance with evacuation orders, as one would hope, varies considerably depending on a household's location: households near the coast and in evacuation zones left at higher rates (the darker areas on Figure 4.1) than those further from the coast or outside the zones. These data suggest that 54 per cent (± 5 per cent) of all households located in an evacuation zone evacuated entirely, with an additional 5 per cent partially complying, i.e. someone in the household remained behind. The proportions were higher in coastal evacuation zones, with 71 per cent (± 11 per cent) evacuating entirely and 2 per cent partially evacuating. By way of comparison, evacuation rates of around 50 per cent have been found to be typical in the face of major disasters (Drabek 1986: 103–5), but rates of around 83 per cent for high-risk areas in major hurricanes (Baker 1991). The Hurricane Andrew rates are suggestive of low compliance for a high-risk area.

For emergency management officials, evacuation actions present two types of problems. The first is households failing to evacuate when ordered to do so – in this case 41 per cent (± 5 per cent) who reported living in evacuation zones did not leave and almost 30 per cent (± 11 per cent) in high-risk beach evacuation zones remained. The second problem is termed "shadow" evacuation, or households from low-risk areas "shadowing" those evacuating from higher risk areas (Baker 1991: 295). Shadow evacuees represent a safety concern because they can add significantly to traffic congestion and increase clearance times. This is particularly the case in South Florida where there are relatively few evacuation routes in the first place and major routes follow along coastal areas. A hurricane heading for South Florida, but turning north before coming onshore, could trap many people in their cars and in more vulnerable positions. Before Hurricane Andrew, about 14 per cent (± 3 per cent) of households outside evacuation zones either completely or partially evacuated. While this rate may seem small, given the population of Dade, it contributed considerably to transportation difficulties. Had the storm come ashore further north, it could also have placed many families at undue risk.

FACTORS INFLUENCING EVACUATION

The factors impelling households to take evacuation orders seriously are no doubt the same ones which led those outside evacuation zones to leave their homes – the belief that a major life-threatening hurricane was approaching. Perry, Lindell and Greene (1980: 151) argue that three factors are critical in the decision-making process: "(1) the definition of the threat as real . . . (2) the level of perceived personal risk . . . and (3) the presence of an adaptive plan." By Sunday afternoon Andrew was recognized as a real threat and many perceived personal risk. For example, when respondents from households that did not evacuate were asked their reasons for not leaving, 91 per cent believed their homes to be safe, i.e. they did not perceive any risk.[3]

While risk is important, other factors also influence the decision to evacuate. Some are likely to be idiosyncratic or unique to a particular household, but it is also likely that other factors are shared across households with similar characteristics and experiences. To capture the relative influence various factors might have on household evacuation, a number of statistical models were developed. Our models were developed on the basis of the following hypotheses derived from the literature.[4]

1 *Households in evacuation areas will evacuate if they are told by authorities to do so.*
This is the most obvious hypothesis and one upon which most emergency management directors depend. Households simply do not have all of the necessary information on which to base an evacuation decision, and yet they live in very vulnerable locations. They must depend on decisions made by authorities and the literature clearly suggests they do. Our sample results substantiate this.

2 *People are more likely to evacuate if they hear the warning in person from a family member, friend, or authority figure rather than just from the media.*
Being told to evacuate in person greatly increases the chances a household will evacuate. We asked our respondents how they found out they were in an evacuation zone and should evacuate – specifically, if they heard it from a friend, neighbor, relative, or authority.

3 *The experience of past hurricanes tends to make people confident they can weather hurricanes in their homes.*
It is often said that experience is the best teacher, i.e. get burned by touching a stove once and you will not do it again. Unfortunately, the hurricane "experience" varies greatly. Hurricanes rarely hit exactly the same place, and they vary in strength, intensity, rainfall, and speed. Nevertheless, experience tends to lessen the probably of evacuating because "hurricane sages" think they could safely go through any hurricane if they have lived through one (Quarantelli 1980: 40). While some South Florida residents had experienced previous hurricanes, few had made evacuation decisions before or had experienced a Category 3+ hurricane. In our data we asked if anyone in the household had experienced a hurricane before and also how long the household had been located in South Florida.

4 *Families headed by aged persons, or extended family households containing aged persons, are less likely to evacuate in response to hazard warnings.*
There are a variety of reasons for this expectation (Perry 1979: 35). First, as

discussed above, older people are more likely to have experienced a hurricane and to believe they can survive in their own homes. Second, the difficulties associated with evacuation, particularly to shelters, are greater for older people. As the young man quoted at the beginning of this chapter said, "My grandmother wanted to stay because she knew that it wouldn't be comfortable being with a lot of people." Third, older people in both urban and rural areas may be more isolated from information about the risks of staying. For our analysis we define elderly household as one in which at least one person was at least 70 years old.

5 *Households with young children are more likely to evacuate.*
Furthermore, women are more likely to plan actively for evacuation while men are more likely to wait passively until ordered to do so. Households with young children can be assumed to be more likely to have the mother as a major decision maker, thus increasing the likelihood of evacuation. In our data, households with at least one child under 10 are designated as households with young children.

6 *Ethnic minorities are less likely to evacuate than Anglos.*
This finding is probably a result of economic conditions rather than race or ethnicity per se in that minorities may have fewer evacuation options. In our analysis, Black and Hispanic evacuation is compared to Anglo evacuation.

7 *People with higher incomes are more able to and, thus, more likely to evacuate.*
They are less constrained by transportation options (such as personal auto, plane, or taxi) and can afford to stay in hotels. To the extent that property security is an issue, higher income neighborhoods tend to be less open to looting and are more likely to have theft insurance; thus, there is less inhibition about leaving.

8 *People living in small households are more likely to be mobile and able to evacuate.*
One of the major findings of evacuation studies is that households tend to evacuate as a whole (Drabek 1983). It follows that smaller households will be less constrained and will find the logistics of evacuation easier. Hence they could be expected to evacuate at higher levels.

9 *People living in multi-unit buildings are more likely to evacuate than those living in single-family dwellings.*
Many multi-family units along the coast are required by management to be evacuated. Furthermore, residents of single-family dwellings, particularly owners, are more likely to be concerned about the security of their property and hence stay to protect.

10 *The ability of a household to evacuate will be contingent on preparation.*
While we do not have a single evaluative measure on level of household preparation, we do have the time a household began to prepare. Given the

relationship between evacuation and preparation time, it is possible that households that were well prepared needed only to start preparation time just prior to evacuating. However, we speculate that in general the later a household began to prepare, the less likely it was to evacuate because it was caught in a dilemma between evacuating versus preparation.

HOUSEHOLD EVACUATION FOR HURRICANE ANDREW

Presented in Table 4.3 are three logistic regression models predicting the log odds that a household evacuated. Controlling for other factors included in each model, the results indicate whether each particular household characteristic had a positive or negative influence on the odds that a household evacuated. Within each model three statistics are given for each hypothesized factor or independent variable. The first is the logistic regression coefficient measuring the unique effect the variable has on household evacuation. If the independent variable increases the odds of evacuation, the coefficient will be positive and significant; if it decreases the odds, it will be negative and significant. The second statistic is a standardized regression coefficient which can be used to compare the relative importance of factors in predicting evacuation. The third statistic, Exp(B), measures the change in the odds of evacuating predicted by the variable. At the bottom of the table are statistics for evaluating the overall prediction success of each model. Easiest to interpret is probably the pseudo R^2 which indicates the proportion of variance in evacuation explained by the model.

The three models predict evacuation for the entire sample using different sets of predictor variables. The first model includes factors generally used to predict evacuation, the second includes only household characteristics and living arrangements, and the third model combines the two sets. While there are some interesting variations among these models, when considered together the factors that have relevance for evacuation are: being located in an evacuation zone, the size of the household, whether it had elder members or children, and residing in a single-family dwelling. As might be expected, being located in an evacuation zone is the single most important factor predicting evacuation. Indeed, being located in an evacuation zone increased the odds of evacuation by over eight times.

Hurricane experience and years in South Florida have significant and negative consequences in the first and second models, but in the third their effects are no longer significant. However, households having elderly members were significantly less likely to evacuate – about one-quarter as likely. Households with children below 10 years old, on the other hand, are about seven times more likely to evacuate. Lastly, households living in single-family housing are about one-third less likely to evacuate.

Overall, three types of variable stand out as unique and significant predictors of evacuation. First, being in an evacuation zone, second, demographic factors

	Model 1	Model 2	Model 3
	Table 4.3 Logistic regression models predicting household evacuation		
Number of cases	1119	1081	929
Constant	−1.45	0.79	−0.35
Aware of living in evacuation zone			
B	1.98[a]		2.12[a]
Standardized B	0.55		0.47
Exp (B)	7.21		8.35
Personally told of evacuation	0.57[b]		0.39
	0.10		0.06
	1.76		1.48
Mean hours before began preparing	−0.01		−0.00
	−0.05		−0.01
	0.10		0.10
Prior hurricane experience	−0.55[a]		−0.36
	−0.16		−0.08
	0.58		0.70
Household size		−0.50[a]	−0.52[a]
		−0.45	−0.38
		0.99	0.60
Years lived in South Florida		−0.01[c]	−0.00
		−0.12	−0.01
		0.99	0.10
Black		−0.07[b]	−0.52
		−0.15	−0.09
		0.99	0.60
Hispanic		−0.39[c]	0.03
		−0.12	0.01
		0.68	1.03
Household income		0.00	0.00
		0.08	0.05
		1.00	1.00
Household with elderly		−1.40[a]	−1.42[a]
		−0.30	−0.25
		0.25	0.24
Household with children		1.64[a]	1.93[a]
		0.44	0.43
		5.14	6.90
Single family dwelling		0.97[a]	−1.01[a]
		−0.28	−0.24
		0.38	0.36
χ^2	218.84[a]	191.68[a]	306.09[a]
pseudo R^2	0.21	0.25	0.34

Source: FIU Hurricane Andrew Survey
Notes: a p≤.01
    b p≤.02
    c p≤.05

associated with having a small household, the presence of either elders (a negative effect) or children (a positive effect), and third, living in a single-family dwelling. It is interesting to note that having children in the household and being located in an evacuation zone closely rival each other in their relative importance as predictors.

Of all the factors considered, the one most often mentioned, and the strongest predictor, is living within an evacuation zone. This factor in some sense captures risk, in that households located in evacuation zones are officially recognized as being in danger of damage and mortality. In light of the salience of this factor, it seems reasonable to suggest that the decision-making process itself, and hence the way factors influence evacuation, would be shaped by whether or not a household is located in an evacuation zone. When we tested how the factors influenced household evacuation depending on whether or not a household was located in an evacuation zone, we found very different processes at work.

The two models in Table 4.4 predict household evacuation separately for those in and outside evacuation zones. While there are similarities with respect to the factors found important before, a number of differences are also evident. First, household size, the presence of an elder or children, and residing in a single-family dwelling remain strong predictors – regardless of a household's location. However, for households within evacuation zones, years in South Florida and income become significant factors as well. In particular, the longer a household has resided in South Florida, the lower its odds for evacuating, while upper income households were much more likely to evacuate. It is also interesting to note that Black households, the majority of which are African-American, residing in evacuation zones were less likely to evacuate, with their odds being reduced by almost two-thirds compared to Anglos. In the case of both models, there is much left to account for when predicting evacuation; however, these findings clearly suggest that very different factors influenced the evacuation decisions made by South Floridians living in these two locations.

SOME POLICY CONSIDERATIONS

These findings are applicable to two problematic areas. Having split the sample into those inside and outside evacuation zones, we can better address the two problems of evacuation: households failing to evacuate when in an evacuation zone and "shadow evacuators" – those living outside zones who nevertheless evacuate. We will consider each separately.

Households in evacuation zones who do not leave

- Complacency appears to be a problem. Simply having lived in the area for a number of years and "experienced" hurricanes does not mean you will survive a major hit.

Table 4.4 Logistic regression models predicting household evacuation, in and out of evacuation zones

	Live in evacuation zone	Not in evacuation zone
Number of cases	372	759
Constant	1.53	0.04
Personally told of evacuation		
B	3.38	0.63
Standardized B	0.05	0.09
Exp(B)	1.46	1.88
Mean hours before began preparing	−0.01	0.00
	−0.03	0.38
	0.99	1.00
Prior hurricane experience	0.03	−0.27
	0.01	−0.09
	1.03	0.77
Household size	−0.38[a]	−0.83[a]
	−0.17	−0.92
	0.69	0.44
Years lived in South Florida	−0.02[c]	0.00
	−0.09	0.03
	0.98	1.00
Black	−0.84[c]	0.47
	−0.08	0.12
	0.43	1.59
Hispanic	0.11	0.20
	0.02	0.07
	1.12	1.22
Household income	0.00[c]	−0.00
	0.08	−0.04
	1.00	1.00
Household with elderly	−1.57[a]	−1.57[b]
	−0.14	−0.41
	0.21	0.21
Household with children	1.88[a]	2.14[a]
	0.28	0.71
	6.56	8.47
Single family dwelling	−1.08[a]	−0.79[a]
	−0.16	−0.28
	0.34	0.46
χ^2	93.02[a]	96.13[a]
pseudo R^2	0.25	0.15

Source: FIU Hurricane Andrew Survey
Notes: a p≤.01
　　　b p≤.02
　　　c p≤.05

- Much more needs to be done to deal with the evacuation of the elderly. Registration efforts have been undertaken in some areas; however, other factors may need to be considered to increase registration to enhance contact with the elderly in times of emergency. The resistance to evacuation centers needs to be addressed – elderly are uncertain about what they will encounter at a center. Perhaps pre-season visits can allay fears.
- Residents of single-family homes need to be made aware that their level of risk is quite similar to that of multi-family unit residents. Indeed, given flooding, they may even be at greater risk. Security issues may have to be dealt with to increase compliance to evacuation orders.
- The problems of low-income and Black households require further exploration. Our results clearly suggest these households were less likely to comply with evacuation orders. We do not know much about the nature of the non-compliance. Likely reasons are lack of transportation and affordable places of refuge.
- Household size appears to be a particularly important factor, in that progressively larger households experience a significant reduction in their odds of evacuating. The problems may be the logistics of coordinating the evacuation of members of large households at the same time, delayed decisions because all household members are not present, or simply the higher costs. Regardless, areas with high densities and large households within evacuation zones should be targeted for educational programs and evacuation incentives.

Households outside evacuation zones who leave

To reduce shadow evacuation, it would appear that the two target populations should be households with small children and households residing in multi-family residences.

- Households with small children must be made aware of the potential harm to their children if they evacuate and are caught out on the road. This might take the form of public notices and school programs. To the extent that households are going to safer locations in the homes of friends or family within the area, timing may be the key element so as not to be caught in transit.
- For households in multi-family units, they too may simply have been evacuating to safer locations with friends and family. However, if they are leaving because they consider their buildings (particularly high-rises) unsafe, the decision to leave may be perfectly rational. Clearly, shelters within large complexes would solve the problem. Short of that, perhaps families can be paired with residents in safer locations within the complex.

These are but a few of the factors that need to be addressed to deal with the two kinds of evacuation problems. Much more needs to be understood before definitive statements can be made. What is clear is that the evacuation rates from evacuation zones were not as high as they should be. Had Andrew hit the most vulnerable areas along the coast of Dade County, more lives would likely have been lost. We were lucky this time; we may not be as fortunate next time.

WHAT PEOPLE WILL DO THE NEXT TIME

The data on Hurricane Andrew evacuation reflects what people really did when faced with an approaching hurricane. It is hard to predict evacuation behavior for a future hypothetical event. Nonetheless, we asked – and found some interesting patterns. Our survey respondents were asked, "If another hurricane were to approach South Florida, would you most likely stay in your home, go to a safer building or structure in the area, leave the area entirely to get out of the path of the storm, or would it depend on the strength of the hurricane?" These data are presented in Figure 4.2, once again mapped by ZIP Code area. The first number is the number of people who say they would leave the area entirely and the second is the number of people who would stay in the area.

While the individual areas fluctuate, there is a clear pattern, reflected by the shading. Households that were in the direct path of Andrew and were subject to hurricane damage were much more likely to say they will leave the area next time. On the other hand, households along the high-risk coastal areas show an even greater likelihood to stay put and ignore evacuation orders. These findings suggest an emergency manager's nightmare: low evacuation for high-risk areas and high evacuation for lower risk areas. Fortunately, at least one of these nightmares did not materialize when the next storm approached.

In July 1995, when Tropical Storm Erin threatened the southeastern Florida coast, the results were both gratifying and potentially alarming. On the positive side, according to media reports, a far greater percentage of homes had properly secured their windows than had been the case for Andrew. South Floridians appear ready to think about preparation in terms of mitigation. Furthermore, the feared mass exodus did not occur. The vast majority of households from the Hurricane Andrew impact area boarded their homes, brought in supplies, and stayed put. Unfortunately, this also occurred within evacuation zones where, despite official orders to evacuate, only a very small percentage chose to leave their homes. Possible explanations include confusion about the orders, failure of evacuation centers to open in many areas, and the fact that it was the beginning of the work week. The most plausible, and hopeful, explanation is that Erin did not strengthen as it drew near. People were closely watching the news reports which painted a decreasingly threatening scenario and, therefore, they did not leave. However, the question remains – if a storm such as Andrew threatened the highly populated and vulnerable areas, would people overcome the complacency of past "experience" and act on the basis of official orders? Emergency managers, meteorologists, and the media must be vigilant in assessing and modeling this potential.

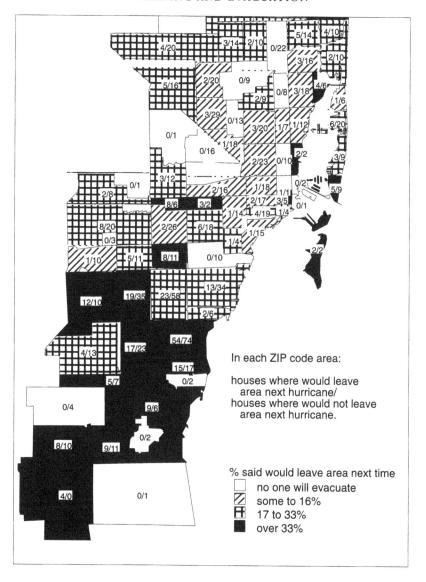

Figure 4.2 Evacuation plans for next hurricane: Dade County households interviewed by ZIP Code

NOTES

1 The data utilized throughout this chapter were collected as part of the FIU Hurricane Andrew Survey. While more detailed discussion of these data are described elsewhere (see Appendix), it is important to note that the unit of analysis is the household rather than the individual.

73

2 Location in an evacuation zone is based upon the self-report of the household respondent. The coastal evacuation zone designation was created by categorizing households according to whether their ZIP Codes fell completely in a coastal evacuation zone. For ZIP Codes split by the evacuation zone, households were included only if they reported being in an evacuation zone.

3 Other answers were nowhere to go (4.2 per cent), pets (3.6 per cent), fear of looting (3.0 per cent), no time to leave (2.0 per cent), physically or mentally unable (1.3 per cent), shelter problem (1.7 per cent), and no transportation (0.7 per cent).

4 For a review of the literature related to evacuation decisions, see Drabek (1986: 100–131).

5

CRISIS DECISION MAKING AND MANAGEMENT

Harvey Averch and Milan J. Dluhy

Where the hell is the Cavalry on this one? We need food. We need water. We need people. For God's sake, where are they?
Kate Hale, Dade County Director of Emergency Management
(*Miami Herald*, 28 August 1992)

The emergency management chaos that occurred in the wake of Hurricane Andrew resulted in a host of relatively standard public administration recommendations (NAPA 1993), including improved planning systems, better coordination, clearer "trigger" mechanisms for federal and state action, and specification of more measurable objectives for emergency managers. Even if they do all these things, however, in an actual event political leaders and emergency managers still must be able to cope with many unpleasant surprises; for example, the inability to obtain the information that *ex ante* plans require. If unable to improvise and react well to a crisis as it unfolds, they can make the crisis more intense by inducing confused decisions or counterproductive behavior. In fact, our analysis of crisis decision making supports a growing body of research that emphasizes the "intergovernmental paradox of emergency management." *Local government is that level of government least likely to perceive emergency management as a key priority and to pay sustained attention to it, but it is handed more and more responsibility for handling disasters* (Cigler 1984; Schneider 1993; Schneider 1992; Wolensky and Wolensky 1990). In the case of Andrew, all federal, state, and city plans called for Metropolitan Dade County to be the premier agent of information and collective action during the initial crisis phase, which we define as lasting six days – the two days before the storm, the day of the storm, and the three aftermath days before the military was called in.

In hindsight, given the design of the intergovernmental response system and the tacit incentives built into it, it is easy to see that it had an inherent, preexisting propensity to break down. Given Andrew's magnitude, it was not surprising that local government did not have the capability to cope. This might have been forecast. For example, in its 1990 study of government response to Hurricane Hugo and the Loma Prieta earthquake, FEMA reported generic difficulties encountered by state and local governments in assessing damage, allocating available resources effectively, and forming appropriate interfaces with federal and state decision makers (FEMA 1990). Thus, decision makers had sufficient prior information to expect, if not predict, that Dade County's emergency management system would be no exception, especially given the magnitude of the disaster. Although Metro Dade initially signaled that it could and would handle the crisis and would not relinquish command to state or federal disaster agencies or to ad hoc trouble shooters sent by the Governor or the President, three days after the storm the situation on the ground forced the county to support the Governor's request for regular military troops. Despite the existence of plans and preparations to make the county government the nerve center for intelligence on damage and for resource allocation, coping with the crisis phase was clearly beyond its capacity (FEMA 1993a; FEMA 1993b; GAO 1993; NAPA 1993).

To illustrate the dimensions of the "intergovernmental paradox" during Hurricane Andrew, we analyze these dimensions that contributed to the failure to manage the crisis well: (1) irreducible natural and technical uncertainties in the warning system; (2) the design of planning at federal, state, county, and local levels as it relates to intergovernmental response; (3) the experience, training, and mind set of the high-level decision makers and the operating emergency managers; and (4) the incentives for voluntary collective action. Then, (5) given the systems and incentives, we describe the decision making styles and strategies of key actors in the crisis and explain why the crisis eventually had to be "solved" by calling in the military. We close by discussing some of the major strategic lessons Andrew taught us and present some suggestions for further research on intergovernmental response systems.

RECONSTRUCTING THE CRISIS

To carry out this analysis twenty-seven open-ended interviews were conducted with key decision makers, actors, and agents at different government levels from April through August 1993. We used a conversational approach to elicit information.[1] Twelve months after the event, most of our informants were still reliving the hurricane response and formulating judgments about their own performance and that of others during the crisis stage. We later circulated a draft analysis to all those interviewed to get feedback. Some respondents provided extensive written comments and were interviewed again at their request. In other cases, we decided on our own to interview again. We were able to check much of the interview data against primary documentary material, such as *ex ante* plans and *ex post* working notes, situation reports, revisions of plans, and "lessons learned" documents from different levels of government as of December 1993 (e.g. FEMA 1993b; GAO 1993; Institute of Government 1993).

CHRONOLOGY OF CRITICAL EVENTS

In examining crisis decision making, it is important to understand the triggering events and the initial environment, the planned structure of command-control-communications, and the designed communication links. These can then be compared to the command-control structure and the communications network that actually emerged in the crisis phase. The chronological structure of the Andrew crisis, along with notes on historical events, is laid out in Table 5.1. It provides the context for our subsequent analysis of decision making and emergency management.

Few officials could reconstruct Table 5.1's chronology accurately during the interviews. In fact, there were many errors in dates and times, although the event itself was vividly remembered. Decision makers often lapsed into anecdotes about their behavior, but these were not always true when cross checked. What was actually happening to decision makers and emergency managers was not clear to them, suggesting that the system suffered wholesale breakdown. However, FEMA's Situation Reports 1–15 served as a cross-check for the chronology we follow (FEMA 1992b).

Warning and intergovernmental response

Emergency plans and actions rely solely or heavily on watches and warnings delivered by the National Hurricane Center (NHC). If these watches and warnings are early and imprecise about location and magnitude, decision makers will act erroneously and incur major social and economic costs. Every false warning, if acted on, costs $50 million and reduces the credibility of future forecasts (Sheets 1990). Mobilizing scarce resources to the wrong areas or evacuating areas that are not hit are both economically, politically, and bureaucratically costly. If predictions are precise about location and magnitude, but come late, jurisdictions may not have time to implement their plans or evacuate their populations. Understandably, any given jurisdiction wants to receive early and accurate watches and warnings. But, collectively, early watches and warnings are not as accurate and, therefore, acting on them can be costly and potentially dangerous. For example, evacuees from a jurisdiction that turns out to be unthreatened can get in the way of the evacuation of those really at risk. Indeed, since the physics of hurricanes allows a significant potential for sudden surprises no matter how good our forecasting procedures, it will be technically and bureaucratically prudent to give warnings as late as possible, to increase the probability that the correct populations and jurisdictions respond.

The paradoxes of warning emerged in the call on Hurricane Andrew. As Andrew approached the east coast, the NHC estimated that it was equally likely to hit areas up and down Florida's east coast. For example, at 5 p.m. on Friday, 21 August the probabilities that Andrew would strike a major urban area within 72 hours along the 600-mile strip from Charleston to Marathon in the Florida

Table 5.1 Chronology of events

Date	Event	Situation	Actual behavior
Friday, 21 August 1992	Emergency Operations Center (EOC) functioning in Dade.	Dade County emergency plan calls for the county manager to take direct command of emergency operations. However, he remains the primary coordinator between the county, municipalities, and other disaster agencies. Plans assume adequate information available at local levels. Coordination assumed because of prior agreement to plan. System not tested in recent times. Assumes needs and information will flow "bottom up" from affected local governments to EOC to state and federal government.	County manager did not take direct command or coordinate, but delegated to Dade County emergency manager.
Saturday, 22 August 1992	Noon: Hurricane watch by National Oceanic and Atmospheric Administration (NOAA). Media begins around-the-clock coverage. Evening: Florida Keys evacuate.	Weather information and analysis pooled into a single forecast by experts in the National Hurricane Center. High uncertainty in forecast until Sunday, or 24 hours before landfall (Sheets 1993).	Probability of a strike in South Dade not fixed until Sunday. County presses NHC for earlier warning.
Sunday, 23 August 1992	Morning: Official warning by NHC. Plans activated by local governments. Dade County activates, but 46 shelters not ready. Afternoon: FEMA representative goes to Tallahassee to be with Governor.	Actors and agents had quite different objectives and responsibilities and information requirements.	Evacuation went smoothly. No collective decision making procedures emerge.

Table 5.1 cont.,

Date	Event	Situation	Actual behavior
	Evening: Request for Federal Assistance prepared and ready to send to the President. Dade County EOC loaded with agents from all 26 jurisdictions, Dade critical department heads, county manager, Red Cross, Governor's representative.		There is general information not keyed to particular organizational requirements. Most agencies unable to do damage assessment relevant to their decisions and to show they acted accountably afterward when audited. There is chaos at the EOC. Lines of command and control unclear. Telephone lines jammed. Dade County requests for help too general and vague for response by higher level emergency managers.
Monday, 24 August 1992	Hurricane landfall at 5 a.m. Before noon: Presidential declaration. President, Governor, and Senator Graham fly over Coconut Grove and down to 152nd Street, but do not go far enough to see major damage in South Dade. President returns to Washington. Noon: Media reporting from the air providing gross damage assessment. Most of day: Jurisdictions begin their own on-the-ground assessments as provided in their plans.	Media shows large swathes of damage. Needs of population inferential.	
Tuesday, 25 August 1992	Governor with FEMA representative does both air and ground assessment of damage. Evening: Homestead city manager calls President directly and asks for military intervention.	Homestead city of 30,000 people faces demand from 350,000.	Homestead city manager emerges as hero because of defection from EOC and plan. Provides more incentives for defection. Florida City defects.

79

Table 5.1 cont.,

Date	Event	Situation	Actual behavior
Wednesday, 26 August 1992	High level military officers arrive in Homestead and carry out assessment as requested by Homestead city manager. President puts troops from Ft. Drum, NY on mobilization alert.	EOC still chaotic with little formal command and control.	Lt. Governor and state officials arrive in EOC to get direct information and provide direct coordination.
Thursday, 27 August 1992	Noon: At press conference, Kate Hale asks "Where the hell is the Cavalry?" Formal FEMA field teams arrive and complete initial damage assessment in the afternoon. Governor Chiles makes formal written request for the military to President Bush.	Dade County emergency manager steps out of channels and admits help is needed. FEMA needs to do its own damage (needs) assessments before acting.	Emergency manager seen to be a substitute for absent or weak political leadership, although Dade County emergency manager reluctant to yield control until Thursday noon. After the emergency manager calls for the military, FEMA director, Lt. Governor, and presidential liaison arrive at the EOC.
Friday, 28 August 1992	Power and telephone crews begin to restore service. A military public information team arrives and sets up operations with FEMA at the airport. FEMA and military refuse to locate at the County EOC.	Florida Power and Light calls on utilities from southeastern states according to prearrangments. Military starts providing basic services.	Private utilities had prior experience in restoring communications and power. Utility companies practiced pooling resources and trained field repair crews.
Saturday, 29 August 1992	Federal troops begin deployment. 23,000 eventually needed.		Large pools of resources now under central command.

Keys was about 4 per cent for Charleston, South Carolina and in Florida, 8 per cent for Fort Pierce, 8 per cent for West Palm Beach, 7 per cent for Miami, and 6 per cent in the Keys (Sheets 1993b). Given the lack of precision, it was not prudent to mobilize. Most of these places would, indeed, have the pleasant weekend predicted by NHC's chief forecaster. At 5 p.m on Saturday the NHC issued a watch for southeast Florida. At that time, there was still high uncertainty, with probabilities ranging from 16 per cent for Cocoa Beach, 21 per cent for Miami, to 19 per cent for Marathon in the Keys. Thus, an early warning could easily have been a false alarm, mobilizing the wrong decision makers and populations and causing resources to flow in the wrong direction. The final warning on 23 August still extended to above Ft. Pierce, about 100 miles north of Dade County. Pundits might say that the definitive statement was the radio report that a massive blast of wind had toppled the NHC communications tower in Coral Gables early Monday morning.

Given that it is not possible today to predict small variations in hurricane tracks in real time, and probably never will be, forecasters have incentives to call "late" when they have the most information about where a hurricane is really going. The NHC is bound, technically and bureaucratically, by the probabilities generated by its forecasting models and the expert judgments of its forecasters. Emergency managers, of course, want *optimal* time to prepare, neither too early or too late. There will always be some derived uncertainty about when plans should be activated and whose plans they should be. In this case, local emergency managers believed the NHC was being too conservative and report that they argued for an earlier warning.[2] On the weekend the public began to act on the more vivid mass media interpretations, such as those of local weatherman, Bryan Norcross, who was warning that Andrew might be the "big one" long overdue in South Florida and that necessary precautions should be taken. And even though coastal evacuation costs $1 million per coastal mile, Dade and Monroe counties decided to give the evacuation order on Saturday – before the official NHC warning.

INTERGOVERNMENTAL PLANNING

Ex ante plans set the context for emergency management by each agent and actor in the intergovernmental response system. For Andrew, the relevant plans were the Dade County 1992 Emergency Operations plan, the Homestead plan, the Florida City plan, and FEMA's general plans (Metro Dade Office of Emergency Management 1992; City of Homestead 1992; FEMA 1992a). Reading across the plans, the most striking thing about them is that none discuss intergovernmental cooperation, nor what to do if communications were blacked out and information sparse or nonexistent. In other words, the plans assume that everyone will have decision-relevant information and that cooperation will be forthcoming by all. The Metro Dade Emergency Operations Center

(EOC) was the designated coordinating body, but the people running it were never envisioned as political agents or negotiators. Thus, managing a hurricane was seen as a technical coordination problem among cooperating jurisdictions, not a political problem of bargaining and negotiation between competing jurisdictions, all with low information states.

Thus, each plan was concerned with the "internalities" of response, not the externalities and relations certain to arise. For example, the Homestead plan instructs the specific functional departments in the city on what they are supposed to do and when. The Florida City plan was structurally similar. The Dade County plan also laid out internal responsibilities. It did not discuss intergovernmental cooperation, although representatives from key Metro agencies and the independent jurisdictions were to be collocated in the EOC to coordinate requests for assistance. Ironically, the plan provided for no formal representation in the EOC for the state Division of Emergency Management or FEMA (Metro Dade 1992).

Natural disasters are explicit political events (Rosenthal and t'Hart 1991; t'Hart 1993). Elected officials had no formal role in the Dade plan. Tacitly, however, it required the appointed county manager and other operating officials selected for their technocratic and managerial skills to become heroic political leaders, carrying out the necessary jobs of public reassurance, networking, and coalition building. Indeed, Lieutenant Governor Buddy McKay later had to become the chief coordinator and intergovernmental liaison operating full-time out of the EOC – a role not envisioned or mentioned in the plan. He functioned as the principal political liaison between local, state, and federal government. The state also placed an official in the Homestead city hall on a full-time basis to do trouble shooting.

The FEMA Federal Response Plan in effect at the time was primarily concerned with coordinating federal agencies to assist states in disasters. In theory, states receive information and requests for assistance from localities which they then relay to FEMA (FEMA 1992b). FEMA then makes detailed needs assessments and allocates resources accordingly. All of our FEMA respondents stressed the importance of timely, detailed needs assessments. They made a distinction between damage assessment and needs assessment. Applying a simple decision rule of sending everything to the site of Andrew based on general damage reports was never an option, partly because FEMA has to be ready for more than one disaster at a time and partly because, like any federal agency, FEMA is held accountable for any wasteful expenditure of funds.

Largely because FEMA became the principal villain, as seen by the media, the agency quickly moved to Miami and placed an officer in the Dade County EOC to directly transmit requests and to keep records on their progress. At the beginning of the crisis, it is clear that the disaster contingency plans, as designed, were not working. Decision makers had to intervene to get information and resources flowing.

EXPERIENCE AND MANAGEMENT STYLE

With very few exceptions, emergency managers at all levels of government in Dade County had never directly experienced any type of hurricane, let alone a Category 4 one. The last hurricane of that magnitude to hit South Florida was Hurricane Donna in 1960 (Hebert, Jarrell and Mayfield 1993). Prior experience might have conditioned them to expect the inevitable surprises and necessity to improvise. In war, commanders find that the first things to go out the window are the peace-time war plans. But the process of peace-time planning presumably teaches commanders to handle the actual situation that emerges on the ground or in the air. However, emergency management training routinizes the plan and teaches managers to stay within its parameters.[3] Instead of preparing decision makers and managers to think about the unthinkable, emergency planning calls for them to stick rigidly to the plan. With some very notable exceptions, most of the key actors in the crisis phase relied tenaciously on their emergency plans, their standard operating procedures for decision making and resource allocation, and the intergovernmental command and control structure that was supposed to be working at the EOC.

Most of our respondents expressed distaste for improvisation and going out of channels, even when it proved successful in gaining objectives. "Not my style," was a frequent comment when asked about improvisation and learning by doing. What experienced managers mean by this comment is that they have learned in many standard bureaucratic situations that acting on their own, using immediate experiences and observations, can be very costly to their organizations and their careers, however correct their actions might prove to be. Thus, the public plea for federal help by Kate Hale, Dade County's Director of Emergency Management, quoted at the beginning of this chapter, was viewed distastefully by most of our respondents. Her action received intense criticism from some Dade County commissioners and her political superiors; she was even accused of embarrassing the President. Nevertheless, someone in Dade County needed to define what the crisis was all about and what strategies were appropriate and necessary to cope with it (t'Hart 1993). In this sense, her actions, while clearly with political ramifications, appeared to facilitate the mobilization of the military.

As it became clearer that original plans and local action were inadequate, other "Lone Rangers" emerged – heroes to the media and the public. They took responsibility and actions that went beyond their job description, went outside accepted political channels, violated bureaucratic codes, or took risks that others chose not to take.[4] These individuals became the "situational leaders" of the crisis phase. In the same situation, many others complained about peers and counterparts or passed the buck. Indeed, blame and finger-pointing continue to this day.

In such situations, we suggest that flexible, adaptive people who learn fast are required. Emergency planners have never been selected with these skills in mind nor trained at acquiring them. For the most part, they have been low-key personalities. Indeed, emergency management positions (from FEMA on down) are not considered to be on the fast track to bureaucratic success. A good lesson to learn from Andrew is to train technocrats and political leaders the way they need to react when they encounter unexpected contingencies (Weisman and Moore 1993).

INDIVIDUAL VERSUS COLLECTIVE RATIONALITY

For decision makers, the intense crisis phase could be likened to a real-life prisoners' dilemma. Each government jurisdiction had an incentive to defect from the *ex ante* central plan and to make its own ad hoc, one-on-one deals for assistance. Dade County emergency managers had no way to enforce cooperation, even though the county might have been better off with some "rational" central allocation of emergency resources. Many respondents testified that organizational chaos and weak command and control were the characteristic mode of the EOC during the crisis period. Dade's Emergency Operations Center was built in 1956 for a maximum occupancy of about 100 people. During the crisis, there were about 300 people in the facility, including the press. According to respondents, people would literally bump into each other, find something in common, and strike one-on-one deals. Central decision makers could maintain no credibility in the rationality of their decisions and actions. They were initially without persuasive technical and political information. Their main responsibility was perceived to be restricted to the unincorporated areas of Dade County. There was little prior tradition of cooperation and successful working relationships with local jurisdictions. Game theory tells us that incentives to defect in prisoners' dilemma type of situations can sometimes be overcome by prior histories of cooperation and regard. Such a history was not present in Dade County, Florida at the time of Hurricane Andrew.

Politically independent, geographically and psychologically distant communities, such as Homestead and Florida City, had no trouble perceiving that their needs would have second priority. Their agents in the EOC reported this perception back to their political and bureaucratic superiors. The latter were, of course, accountable to their immediate voters. As a high-level decision maker in one municipality said, "Once we saw what was happening at the EOC, we told our representative to come home and take care of the people in our city." The hoarding of resources and going outside regular channels to acquire more became natural and understandable responses to the breakdown in the Dade County command and control system.

As the inadequacy of the county response became more public, county-level officials remained highly reluctant to relinquish control. Finally, it was clear to all that regular army troops in large numbers would be necessary. Exactly when this became clear, and when the requests were made, remain a matter of dispute among those we interviewed. We suspect everyone did a little "bolstering," now emphasizing their prescience (Calvert 1985).

The EOC was the only vehicle designed for collective group action. The consensus in our interviews is that it failed badly in the three days after landfall. In fact, it was bound to fail, because its design assumed that there would be voluntary coordination. This might well have happened if damage had been light. Thus, the costs to a particular jurisdiction of pooling resources and allowing central emergency managers to make allocations would have been relatively low. In the case of damage this widespread and severe, however, the costs of letting resources go and of not competing for additional ones were inordinately high. The role of the EOC director, as originally designed, was to be a persuader and coordinator, not a commander with the ability to allocate scarce resources. The appointed county manager was supposed to take command under the plan, but he chose not to, or felt he could not. The assumption that the disparate interests collected in the EOC would suddenly, after years of distrust, act altruistically in a severe situation without political leadership proved untenable. Thus, central emergency managers could not enforce compliance with whatever decisions they believed benefited the county as a whole.

The Dade County Manager might have assumed direct control in the EOC and made a difference. His authority is greater than that of his emergency department head, the officially designated manager of the EOC. However, respondents confirm that his operational style was to delegate to those responsible and in possession of the necessary information – a style which frustrated a number of senior county staff. According to our respondents, the county manager made no effort to collect the principal actors in one place for consultation as the crisis developed. Instead he relied on one-on-one communications, using the ordinary chains of command. He did not hold a formal staff meeting with his department directors until 2½ days after the hurricane.

In ordinary bureaucratic situations, management literature suggests decentralized, hands-off decision making is a good strategy. High level leaders set strategic paths and select appropriate managers. But the use of this standard strategy in an unfolding crisis becomes problematic. The crisis management literature suggests that in intense crises decisions should be made at the top of the organization because those at lower levels tend to "suboptimize" based on their own interests, do not have all the information necessary, and are unaware of larger political or social issues or constraints. Thus, for example, the need for an executive committee in the Cuban missile crisis. Previous research has documented the emergence of "synthetic groups" in a number of urban disasters (Wolensky and Wolensky 1990). A synthetic group – a working coalition of key actors and agents at different levels – might have been able to resolve some of the command and control problems and to facilitate damage assessment and resource deployment (Dluhy 1990). However, no such group or coalition formed during the crisis period.[5]

HOW THE DECISION MAKING SYSTEM OPERATED

Once warning was given, the planned intergovernmental response system was simple in design. Collective decisions would emerge from the EOC, to be implemented by county and local operating agencies. Demands for federal resources would flow from the county to the state to FEMA. However, the interaction patterns which evolved during Andrew were complicated. Visualize a bicycle wheel with spokes radiating out from the center. The NHC is located at the center of the wheel, providing initial information – watches and warnings – that then radiate along the spokes. For technical, economic, and bureaucratic reasons, the NHC has incentives to make warnings relatively late but with high accuracy. The NHC center of the wheel was partially displaced by the unexpected emergence of a media actor, Bryan Norcross, the Channel 4 meteorologist, who made his unofficial warning call 24 hours before the official one. He used NHC's information, but was freer to interpret it. As someone in the private sector, the potential costs to him of an early, perhaps unnecessary call were less than they were for the NHC.

On the wheel, spinning around the now two overlapping centers are the emergency managers from the different governmental jurisdictions. They are trained in the routines of preparedness and recovery. They attend numerous seminars and workshops and become familiar with plans and operating procedures. They follow orders, stay within the designated command and control structure, and implement the emergency plans as written. They are not taught, nor are they expected, to improvise. Until the NHC makes a call, their main job is to prepare, then wait. It is unlikely that they would follow Norcross' message rather than the NHC's and decide to act early and mobilize on their own.

Managers, including those from the American Red Cross, have a critical additional function to perform – to inform, mobilize, and connect the elected and appointed public officials who have the authority and responsibility to act. Emergency managers are not trained in diplomatic and political skills, even though a few exhibited these traits. With a few exceptions, the technocrats adhered to the *ex ante* command and control structure. They tried to follow the standard operating procedures detailed in their emergency plans. This compliance with set plans may work in disasters of a lesser magnitude, but we believe that in a major disaster different behavior needs to be instilled.

There was a notable exception among the emergency managers – Kate Hale. While she did not challenge the timing of NHC's warning and, for the most part, followed the Dade plan after the hurricane was over, she did step outside her job and normal technocratic role. When she challenged the federal government and President Bush on Thursday, she became the symbolic political leader of a leaderless Dade County. Technocrats are not supposed to coerce the President or higher authorities or manage a disaster from a political perspective. They are supposed to play by the rules and work inside the system. In the short run, Hale's framing of the crisis by calling for the military made her a local, even national, hero for bucking the system. But, as indicated earlier, her action reinforced the traditional behavior in Dade of blaming other organizations for failure. In standard bureaucracies, those who behave as she did suffer retaliation or punishment when the opportunity arises, regardless of the social merit of their action. And, in point of fact, Kate Hale is no longer Dade County's emergency manager.

Spinning around the technocrats who know the plans and collect information are high-level elected and appointed public officials. They, typically, are not experts on emergency management, but have political responsibility for decisions. These are the city managers, department heads, mayors, commissioners and council persons, and the federal and state officials responsible for resource acquisition or deployment. In our interviews of a cross-section of these officials, we attempted to understand the behavior and decision making processes of each of our informants.

The actions of Alex Muxo, the city manager of Homestead, resembled Kate Hale in one respect. He improvised and did what he saw as necessary for his jurisdiction. However, unlike Hale, who appealed for help across the board, Muxo, first and foremost, protected the interests of his constituency of 30,000 Homestead residents, although at the time there were 350,000 South Dade residents needing services. The day after the hurricane, Muxo hired a public relations firm for $70,000 to deal with the media. The subsequent national and state attention paid off, eventually leading to substantial resources and help for the city of Homestead (see Chapter 11). On Tuesday evening, 26 August, he called President Bush and asked for the military, and got it. He went outside channels more than once and was anything but complacent. In the short run, his activities paid off for his constituency, but there is usually a long-run price. For all practical purposes, he severed relationships with the EOC. He is not perceived as a team player and the Dade bureaucracy remains angry at him to this day. Homestead continues to enjoy a favored status with everyone except the county. In the future, when the city needs something from the county, it may not receive it. Bureaucracies have long organizational memories about people who were not on the team or who got off to pursue their own or the public interests.

Lt. Governor Buddy McKay and US Senator Bob Graham were also among the improvisers. Neither had any formal role to play in the crisis phase, yet both became essential trouble shooters behind the scenes. McKay went directly to the EOC and tried to perform the brokering and networking necessary to deploy emergency resources effectively. According to one federal official, "He was the only one interested in operations and making the system work . . . He was the 'glue.'" Although McKay did not capture much media attention, he did become one of the few "situational leaders" in the EOC. Senator Graham also improvised by going directly to South Dade where he served as political liaison between the President, the Governor, and some of the elected Dade County officials. As a current senator, former governor and resident of Dade, he knew all the important actors and was able to facilitate the initial damage assessment to help mobilize federal resources. By working behind the scenes and avoiding the media, he helped forge the necessary networks.

While these officials improvised, many others in Dade threw up their hands in desperation and harshly criticized the intergovernmental mechanisms in place. They did not try to make them work better, but instead disappeared from view. They failed to improvise, procrastinated, denied personal responsibility, and avoided exposure to disturbing information. Since no synthetic group emerged during the crisis phase, the situational leadership compensated for its absence and tried to cope until the military arrived.

Points on an ordinary, well-balanced, rolling bicycle wheel do not have control of their own fate. Any given point on such a wheel traces out a curve whose shape is predictable with 100 per cent certainty – a cycloid. But the decision wheel in Hurricane Andrew was not like this. Points on our wheel have personal and bureaucratic interests, and they exert force on its overall direction. These self-aware and self-directed points make the wheel wobble and head in directions no one intended. Wheels do not usually roll by themselves, but have someone or something that gives control and direction. But our decision wheel was propelled by the energy of the points on it. So it is understandable that it took outside forces and energy to bring it under control. The military had sufficient personnel, resources, and organization to stop the random wobbling and direction.

SOME POLICY CONSIDERATIONS

Our analysis of crisis decision making during Andrew suggests some obvious lessons in hindsight.

1 *In a Category 4 or 5 hurricane, the civilian intergovernmental response system is inherently political and contentious. It will tend to breakdown because there is not enough time to carry out the ad hoc negotiations and bargaining necessary for political equilibrium.*

It follows that individuals, organizations, and jurisdictions need to be self-sufficient for three to five days after a hurricane. We can reduce the need for bargains and deals, but this will require abandoning the intergovernmental decision making system we now have, and adopting a more arbitrary and centralized command and control structure. Do we want more and improved intergovernmental coordination or more centralized decision making by fiat? We believe Andrew may have moved the nation closer to a centralized system. But we are hardly ready to abandon the decentralized and complex intergovernmental system, although some are ready to consider more seriously a proactive and assertive role for the President, the military, and FEMA during a major disaster. Centralized, dictatorial systems have a different, but potentially serious set of problems. Clearly, the military cannot be relied on as the preferred system for every crisis, especially crises of lesser magnitude.

Future research needs to identify those conditions that predict when state and local governments will not be capable of managing a disaster phase. The currently proposed strategy of early higher level preemption should be evaluated. We believe that a comparative historical analysis of breakdowns would be an appropriate start.

2 *Corrective measures to improve technocratic and comprehensive management – improved communications, repositioning, clearer standard operating procedures – will only improve the crisis management system marginally.*

Many of our respondents noted that no one could plan for an "Andrew." We have argued that, in addition to investments in damage avoidance and in managerial and logistical improvements, at least equal emphasis should be placed on helping government decision makers and emergency managers become more flexible and responsive to surprising information. Finding cost-effective improvisations is a key requirement. Tinkering with the plans or sharpening the prediction models will not help those who have to cope with the surprises. Either we need to select different types of officials, or we need to change the training and incentives we give those we select. The research question here is whether education and training per se can have a major impact on decision making and, if so, what kinds of training would prove cost-effective. The feasibility of training political leaders and officials in addition to emergency managers deserves examination.

3 *There is a fine tradeoff involved in timing watches and warnings, given the different incentives and interests of the actors and agents involved.*

Given these different incentives and interests, we suggest that there needs to be much closer dialogue – perhaps face-to-face communication – related to the advisories coming from the NHC. The NHC has strong incentives to be conservative. However, relying on the next Bryan Norcross may not have as good a result as occurred during Hurricane Andrew. Research might explore the utility of using experts/consultants to assist responsible officials when they must interface with the NHC.

4 *In major disasters where the key actors and agents working in local government have little or no prior experience, possess no political and bargaining skills, and there has been little prior history of cooperation, the* ex ante *intergovernmental response system will fail.*

In Andrew's case the actual behavior was: going outside the system by defecting; blaming other levels of government for personal failure; and/or hoarding resources and avoiding centralized resource planning and allocation to benefit the common good. This behavior, while rational for individual jurisdictions, made the county's attempts at collective action difficult. It was in direct opposition to the designed plan to act collectively in emergencies. But, there is a long history of non-cooperative intergovernmental relations in Dade County.

In the end, as one perceptive federal official looking at South Florida from afar said:

> We did not realize that there were so many communities involved in the recovery effort. We relied on Dade County . . . but we did not anticipate the squabbling . . . We eventually relied on television for information, not the SOPs [standard operating procedures] outlined in the plans for damage assessment . . . The Dade County officials were saying they could handle this and they were reluctant to say they needed help . . . The hesitancy to request the military caused much pain and suffering.

We heard many tales of intelligent action and heroic acts, but we also heard stories of public officials seeking political and organizational gains. We conclude that when a hurricane of Category 4 magnitude arrives in an area where there are historically uncooperative, highly rivalrous jurisdictions, it is unreasonable to expect planned intergovernmental response systems to work well.

Whether the military should become the standard solution to such cases requires careful study. As public administration and policy analysis professionals, we have to believe that the recommended improvements in logistics and communications, plus improved hands-on training in crisis management, can prevent a breakdown or, in the least, mitigate its effects.

NOTES

1 This was a joint project; the authors are listed alphabetically. We thank Mary McDonald, a Ph.D. student at FIU, who helped conduct the interviews and construct the bibliography. See the Appendix for information about the discussion protocol.
2 The NHC monopoly on technical information and forecasting models probably deters state and local decision makers from investing much in their own forecasting expertise. So they are unable to ask deep questions about the NHC forecasts or to conduct probing discussions about decisions to mobilize. We note that the State of Florida has moved to acquire more forecasting expertise.
3 Siegel (1985) lists twenty-five different skills an emergency manager should have to be able to act "rationally" in an emergency. Nearly all of these have to do with *ex ante* "understanding" of rules and procedures. Coping skills under low information situations and intense pressure are not listed.
4 Schneider (1992), in her study of Hurricane Hugo, argues that decision makers should always stick to their *ex ante* standard operating procedures. Departing from them induces an adverse public response which makes management more difficult. For example, the night before Hugo hit, the Governor set up a new unit in his office which made the job of the designated State Emergency Preparedness Division more difficult. However, given the *ex post* stakes, the surprises, and the political consequences, technical managers should expect political leaders to intervene.
5 After the crisis, the private organization, We Will Rebuild (WWR) claimed it was the synthetic recovery organization. WWR originated with a call from President Bush to Alvah Chapman, the historic leader of the Miami business elite, at the end of the crisis period. Strictly defined, WWR was thus not a spontaneously formed, local

synthetic organization. WWR was eventually responsible for the distribution of millions of dollars in private and public funds. According to WWR leaders, it was also instrumental in acquiring additional federal resources and favorable treatment from Congress and the Florida legislature.

6

COPING IN A TEMPORARY WAY

The tent cities[1]

Kevin A. Yelvington

In the fall of 1989, Jeannie Smith saw her house destroyed when Hurricane Hugo smashed the Caribbean island of St. Thomas. With her belongings and job gone, she migrated to Miami with her two children to "start a new life," as she put it. When I met Jeannie in 1992, her home had again been destroyed in a hurricane. She was living in a tent city in Homestead, Florida, one of the numerous victims of Hurricane Andrew. Outside, the harsh South Florida sun beat down on the tent relentlessly. Inside, the heat and humidity cause Jeannie to wipe the sweat constantly from her brow. What belongings she had managed to salvage fit under the three cots allocated to her and her children. In their corner of the tent, plastic containers of water, a Styrofoam cooler, and an empty suitcase sat on the plywood floor. Not flinching from the ruckus of a military helicopter landing nearby, Jeannie's eyes focused on some point in the distance as she wondered out loud: "What do I do now?"

This chapter is about the experience of Jeannie Smith and the thousands like her who, for a number of reasons, found themselves living in the tent cities erected to provide temporary emergency housing in the wake of Hurricane Andrew. I provide the reader with a general sense of what the tent cities were like from the viewpoint of their residents. I also account, at least obliquely, for the social forces that determined just who ended up in the tent cities. At the same time, this chapter is about the role of tent cities as a part of official disaster response policy. In terms of theory, I use the tent cities case to speak to the larger structural inequalities and sociocultural discontinuities that they came to represent. To accomplish this, I'll start by calling on a time-honored tradition in social anthropology: conflict theory.

At least since Max Gluckman's opposition to the prevailing orthodoxy of the structural-functionalism of the 1940s (for example, Gluckman 1958 [1940–1942]), a significant number of anthropologists have chosen to analyze the conflict inherent in all aspects of social life. Unfortunately, as was argued in Chapter 2, mainstream disaster research has inherited from the structural-functionalists the theoretical assumption of the functionally integrated "community" – perhaps the most clichéd and imprecise term in the social science vocabulary. Community is characterized by a certain equilibrium existing before the social interruptions caused by a disaster. As we will see from the data presented here, this position does not hold in the case of Hurricane Andrew's tent cities.

Not only do the micro-negotiations that encompass ethnic, class, and gender identities reveal schisms at the structural level, the face-to-face relations recorded in the ethnographic data below tend to reproduce, but sometimes transform, these larger patterns of social and cultural arrangements. Structural properties simply could not exist without them. One of Gluckman's main points was that anthropologists must take into account the whole of a political situation in their ethnographic fieldwork and in their writing. While discussions of the South Florida context are provided in various chapters throughout this book, I will concern myself with a critical evaluation of the class and cultural assumptions behind the implementation of official policies in order to then provide some recommendations. It is evident that many of the official policies regarding tent cities were far off the cultural mark.

GETTING THERE

As discussed in more detail in Chapter 1, it is now estimated that about 180,000 people were left homeless for some period of time as a result of the destructive forces of Hurricane Andrew. While many had the resources and connections to relocate on their own, many did not. Despite the general perception of a slow federal response, the military quickly responded once the Presidential order was issued (Gladwin 1993: 99–101). A commanding officer described it as "the grandaddy of all assistance operations. We're approaching this like a war – except we're putting troops in the field to help people, not kill them" (Gore 1993: 29). The peak military presence was 23,000 troops and 6,000 members of the National Guard. They began arriving several days after the storm and the last Marine and Army units left in mid-October (Higham 1992b). The military initially established tent cities for homeless victims at three locations: Homestead Middle School, Campbell Drive Elementary School, and Harris Field recreation complex. They opened on 1 September with 108 tents (*Miami Herald* 1992c, 4 September). Later, a fourth camp was opened in Florida City. Designed to provide temporary relief, the tent cities were located in deep South Dade where the physical destruction from Hurricane Andrew was at its greatest. By the time they were dismantled two months later the tent cities had housed nearly 3,500 people.

The relief centers were constructed by the military, who were also in charge of most day-to-day operations, including security and the preparation of meals. The Marines were in charge of the Harris Field site, while various Army units were in charge of the other three. Military security was augmented by the Homestead police force and by US Marshals. In the last weeks of the tent cities existence, these military units were pulled out and replaced by local and federal police.

The tents themselves were military issue and came in two colors: olive drab and the buff-toned "Desert Storm" version. Ventilation was achieved by rolling up their side flaps – at the cost of privacy and of making one's living space mosquito-infested. Built on wooden platforms 25 by 35 feet and designed to accommodate about 18 mosquito-netted cots each, the tents were lit with fluorescent lights which sometimes had faulty electrical wiring. Electric fans were not allowed inside the tents, nor were any alcoholic beverages. All residents were expected to adhere to a 9 p.m. curfew.

The compounds were arranged in rows and columns, with administrative and service tents located in one area (see Figure 6.1). Residential areas were designated for single men, single women, two-parent families – although the residents' definitions of what constituted "family" were often at odds with official definitions, married couples without children, and single mothers with children. Along with a cot and mosquito net, each tent city resident was issued sheets and blankets, a pillow, wash cloths, a towel, and toiletries. Many relief agencies also donated clothing and other items.

As typical of September in South Florida, it was very rainy, hot, and humid. Due to the continuing rain, the compound grounds became so mired in mud that gravel had to be brought in to build dry paths to connect the tents. Lining the perimeter of each compound were dozens of portable toilets while the portable, communal showers were located in common areas. Residents used the showers on a rotating basis, with certain times reserved for women and other times for men.

When first established, American Red Cross volunteers were in charge of checking in victims and a rather flexible policy was adopted in which people were generally given admittance without fully proving that they were living in a dwelling destroyed by the storm. The general feeling was that anyone willing to live under the harsh conditions must be in dire need of housing. Those who arrived and registered as a family were usually placed in the same tent. Our research team came across a number of different family systems organized around various kinship links. For example, Felicia Patterson, a 40-year-old African-American woman from Homestead, explained: "My mother's here with me . . . My sister's deceased, so I have her three kids and my three kids here. Theo's [her eldest son] up to my aunt's . . . Down here there's eight of us."

At first, the flow into the tent cities was only a trickle. Only 41 people were registered at the 1,500-capacity Harris Field facility on 4 September – ten days after the storm – and the Florida City one, built for the same capacity, housed only 58 storm victims (*Miami Herald* 1992d, 5 September). The reasons for this early under-utilization are numerous and some relate to the area's ethnic and class context. Newspaper reports quoted area residents who expressed reluctance and fear of moving into the tents. One resident who reluctantly moved into a tent with his wife and three children wanted us to know, "I've never lived like this before" (*Miami Herald* 1992d, 5 September).

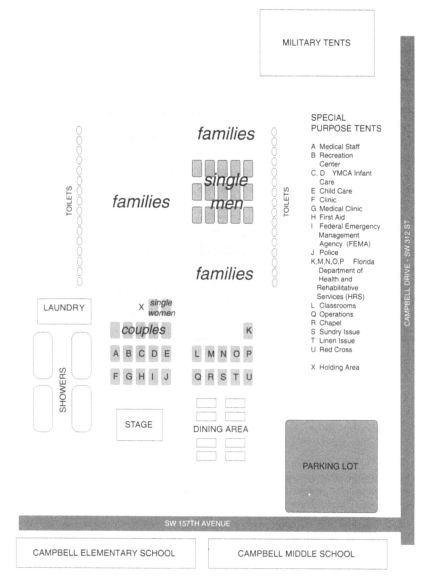

Figure 6.1 Campbell Relief Support Center

Fear of being arrested by immigration officials if they came to a tent city was expressed by agricultural workers, many of whom had previously lived in the trailers and low-income housing provided by the large agribusinesses around Homestead. Authorities subsequently promised that undocumented workers would not be arrested and FEMA asked Immigration and Naturalization Service officials in Washington to stop Border Patrol officers from wearing their uniforms while assisting with food and supply convoys (*Miami Herald* 1992c, 4 September).

Another reason for the under-utilization was that much of the early hurricane relief information was only given in English. The assumption that all hurricane victims were English-speakers prevented the distribution of food, medical supplies, and assistance information to some of the area's Latinos and Haitians. To surmount the language barrier, recovery information was eventually made available in Spanish and Haitian Creole. The military was the last to make recovery information available in languages other than English.

As the post-hurricane days mounted, increasing numbers of homes and dwellings were condemned, other options for shelter disappeared, and more people began moving into the tent cities. There are documented cases of families ordered out of their residences by officials with the enticement of rapid place-ment in a FEMA trailer, only to end up living in their cars (Higham 1992a). In addition, when landlords received their insurance money for repairs, the tenants were often evicted while the reconstruction took place (see Alvarez 1992a). Others came to the tents after county officials evicted them from condemned houses and mobile homes on short notice. Residents from one trailer park actively resisted eviction, stating that they did not want to lose track of their neighbors or leave their belongings to be vandalized. Their efforts failed and the residents were forcibly relocated to the tent cities by US Marshals (*Miami Herald* 1992g, 12 September).

The net effect of all these factors was that the population in the tent cities rose rapidly. On 7 September, the population of the Florida City camp nearly quadrupled – from 120 to 467 residents – and at the Harris Field site the population doubled to 1,125 on the same evening. By the end of September, a month after the hurricane, the population of the four tent cities was reported to be more than 4,000 (*Miami Herald* 1992i, 27 September). At the same time, Centro Campesino, a migrant farmworker advocacy and community develop-ment organization, opened its own tent city in the most southern part of Dade County, housing hundreds of storm victims in tents sent by the Mexican embassy (*Miami Herald* 1992j, 29 September). Nevertheless, a FEMA Recovery After Action Report stated that few disaster families and individuals actually moved into the tent cities and attributed this in large part to poor management, alluding to the fact that the Federal Response Plan called for the military to erect and maintain the facilities, but did not identify the organization to manage them (Polny 1993).

From the onset the tent cities were intended to be a stop-gap measure until better housing alternatives could be found. FEMA immediately began the long process of locating and preparing sites to bring in thousands of trailers and mobile homes. Processing the tremendous number of applications for trailers and other aid was a difficult and time-consuming process. Some victims were coaxed into the tent cities by the perception that they would get help sooner if

Plate 6.1 An elderly woman receives her cot at the Centro Campesino Farmworker's
Association campsite
Source: Al Diaz/*Miami Herald*

they were there. Others came there whose homes were not physically destroyed,
but because they had lost jobs or systems of support. Monica Sánchez, a 17-
year-old Mexican-American, living at Campbell Drive with her mother, four
sisters, two brothers, and a half-sister, is a case in point: "Well, our apartment is
still up . . . only broken windows. But we came over here because we didn't get
no help or nothin'." [Learning that her apartment had not been condemned, the
interviewer said, "You decided to come to the tent city because . . . "] "'cause we
couldn't get no help or nothing there but clothes . . . so my mom went to Red
Cross and told them what happened, so they told us to come over here. It's
better for us. And so we came . . . But, we haven't got our trailer yet, we've been
here three weeks." Still others came to the tent cities after spending time in
one of the numerous shelters across the county. Many of these shelters had
themselves been damaged during the hurricane. All of the usable shelters located
in schools had to be closed in order for the schools to open two weeks after
Andrew.

The tent cities were often the last resort for families who first tried to access
the resources of friends and relatives (see Chapter 8). Many who normally would
have provided assistance with food and shelter, however, were themselves
victims. Not only did almost every tent city resident we interviewed report that

they had no homeowners' or renters' insurance, but many who found themselves in tent cities lacked the aid and support of relatives and friends. Given the general class homogeneity of social and kin networks, most tent city dwellers had to choose from among those who were in similar situations as themselves. In other words, the resources available through the support networks of those who ended up in the tent cities were all but depleted. Many tent city residents knew or were related to each other. Some were sharing tents with relatives with whom they had not previously shared a household. In some cases, tent city dwellers felt they were actually better off than relatives who had not moved there.

BEING THERE

Much of a tent city resident's life was spent standing in line – for bottled water, for food, to take a shower, to be put on a list for jobs, to receive donated clothing, and for different government assistance programs. A number of state and federal relief agencies set up tents in the camps to provide services such as psychological counseling, medical treatment, daycare, and job placement. Other agencies set up after-school programs for the camps' many children. In addition, FEMA established a claims processing site and Southern Bell set up banks of phones, allowing free five-minute long-distance and international calls.

Psychological counseling services appeared to be highly utilized, especially during the first weeks. A number of instances of post-traumatic stress disorder were evident during our interviews. Many camp residents (as well as hurricane victims in general) reported psychological depression and strained familial relations – including an increase in reports of domestic violence and spouse abuse – during the immediate aftermath (Morrow 1993). Some of our respondents attributed the behavior of others to stress caused by the storm and to the lack of support experienced by tent city dwellers.

Most agreed that children had an especially difficult time. Interviewed while sitting on her cot in a tent two weeks after the hurricane, Betty Knowles, a 34-year-old African-American mother of five children ranging in ages from seven months to 15 years, said of the hurricane, "it was something I'll never forget . . . when it rains hard now, we [the entire family] pray so hard . . . Because they think it's the storm again." John Babb, a 43-year-old Anglo truck driver, compared tent city living to his experiences in Vietnam. "It's just about like Vietnam except for the fighting. It is primitive . . . All you gotta do is look around. Look at the way we have to wash our clothes right now to keep 'em clean. It could be worse, though . . . I'm grateful for what we have, but its just . . . a bitch sometimes man."

Finding work became a major task for many. Some of the job placement services were successful in landing residents temporary jobs associated with construction and hurricane debris cleanup. Officials also began hiring residents for a number of tent city jobs, such as cleaning up and patrolling the parking lots. Bud Talley, a 51-year-old unemployed cabinet maker, was asked what he and his tent-mates, all single men, did during the day:

> Well, first, we try and plan for the next day's work . . . cash work . . . anything like that. This fellow here [referring to a tent occupant], he is keeping in contact with his family, this and that. This fellow here [referring to another occupant], he is working for a gardener . . . someplace where they lost a bunch of plants . . . He is working there during the day, getting it together. And myself, I am working around, working on my car, like I did about an hour or so this morning . . . and looking for day work.

He had a strong belief that workers were being exploited.

> I got nothing going on this morning, so I went up there and I said, "What are you doing?" "We're doing roofs." "How much you getting?" "Five dollars an hour." I just told him right in front everybody, "You just taking advantage."

[Interviewer: "At the Red Cross tent there were notices up about different kinds of jobs. You think a lot of people are using those services?"]

> Yes, most of the guys, especially the family guys, are jumping right on it. They got something going . . . they are grabbing something everyday . . . Everybody was throwing the trash out on the grass and everything, you know, and the Marines got to come by and pick it up. It's kind of demeaning . . . to make a military man do it, being a old military man myself. But the military finally put it on the other foot and hired a few guys, paying them cash, and they're going around all day.

For many, not only were the tent cities seen as blessings, but the assistance provided by the battery of agencies in the tent cities was welcome and appreciated. Roberta Campos, a Mexican woman, reported:

> Well, we received [help] when we arrived here and realized it was not as horrible as people said. Since we arrived soldiers passed by every 15 minutes to ask us what we wanted, what we needed, tell us that we will eat or shower at such-and-such time, that food has arrived, or clothing is being given out. We were received since the beginning, very warmly and . . . were treated like people. In addition, the same night that we arrived we went [to] a support group, a psychologist. She let us talk and vent what we had inside.

Liza Jones, a 63-year-old African-American who was living in a tent city with four grandchildren, ages three to 12 years, said:

> I am grateful to be here . . . and the people have been so nice, they have been so nice. I didn't know that people can be this nice to you . . . and I didn't know that I would share this tent with a stranger. That's what I have

been doing for the last . . . how many days since the storm? . . . I have been living on the kindness of strangers.

In the first few days there was a palpable ethos of cooperation and sharing among the residents. This was the cause for much optimism and hope. As Conchita Torres, a 32-year-old Afro-Puerto Rican, said of her experiences in a shelter after the storm:

> I have never seen this before, I never thought this sort of thing would happen . . . I think [it] is beautiful . . . that everybody is getting along, like Spanish and Black. They didn't get along and now in the shelter I see . . . White and Black together hugging. That makes me feel so beautiful. I always wanted this, always. I see people, everyone getting along: Black, White, Spanish, Haitians, everybody . . . they try to help each other.

She said this spirit of cooperation pervaded the tent city as well, although not to the same degree.

When Emilio Barron, a 42-year-old Mexican, was asked about this spirit of cooperation he responded:

> Well, the only thing that I see changed now is that everybody is getting closer to each other. Everybody is talking, everybody is friendly to each other . . . Like my sister in Leisure City. She didn't even know her neighbors, they didn't even talk to her. Now they offer her food, they offer her whatever she needs. Everyone is more friendly with this thing that happened.

However, most tent city residents, including Barron, thought the post-hurricane friendships would be short-lived. "When things get back to normal again, when we all have our house and everything . . . everyone's going to be again the same. That's what I think." Indeed, many said that there had been a change in attitude after the realization of their plight set in and they began dealing with the relief bureaucracies and the unrelenting heat. As one woman put it, "the camaraderie is kind of evaporating now."

These feelings corresponded to a general ethic captured by other research of the FIU disaster team. In a Dade County-wide random telephone survey, 81 per cent said that relations with neighbors and others were better immediately after the hurricane, while 16 per cent felt that these relations were about the same. However, by December 1992, 60 per cent reported that things were back to normal, while 38 per cent said relations were still good. These feelings corresponded to people's actions: in the first two weeks after the storm, there were thousands of volunteers working with a range of aid agencies. However, by the third week the numbers dropped dramatically as the city began to return to a normal business schedule (*Miami Herald* 1992h, 12 September).

In the tent cities, the spirit of cooperation decreased as frustration levels rose and days turned into weeks, then months. People became impatient when

100

FEMA's promised relief was slow in coming. Many told of being frustrated by the lack of personal space and privacy. When they could get them, residents hung blankets and other objects up to "wall off" their area, giving some semblance of privacy within the tents. Frustration sometimes was manifested in verbal altercations and fist-fights, especially between men. Cases of domestic violence and spouse abuse were officially reported or – more commonly – discussed in hushed tones amongst camp residents. The covert presence of illegal drugs and alcohol served to fuel these confrontations, which usually occurred late at night. As John Babb commented:

> You see arguments, you see fights, there's drugs in here, not in this tent, but there's drugs here on the compound. There's alcohol on the camp . . . It's not really much different right here than it is anywhere, [in] any other neighborhood in this whole area or anywhere in the nation. The only difference is that you might have to go . . . pick it up and smuggle it back in. You have families that you'll hear . . . out in the middle of the street . . . fussing and arguing and kids screaming and crying, but . . . it all mellows out. There's been a few in fights, there's been a few caught with drugs. It's no different than being and living over here in Homestead or anywhere else, really. It's like living next door to your neighbor and if the police come one night and take his brother away, you know, for having something and you say, "Gosh, I didn't know those people were doing that." It's the same here . . . It's just like another little small community.

There was considerable ethnic diversity in the tent cities, and many instances of frustration and conflict turned on the issue of ethnicity. While no known census was compiled, we made some tentative "guesstimates" of the ethnic composition. We estimate that tent cities population was between 50–60 per cent Latino, comprised mainly of Mexicans and Mexican-Americans, Central Americans, and Puerto Ricans, in that order. There seemed to be very few Cubans or Cuban-Americans. Blacks comprised about 30 per cent of the total and included African-Americans, Haitians, Haitian-Americans, and other Afro-West Indians. This is not to say that "Black" is an exclusive category. In fact, social divisions amongst the Black populations in South Florida are often far more significant than their similarities.

The remaining 10–20 per cent of tent city residents were non-Hispanic Whites, known locally as Anglos. Many from this category lived in the Homestead area before the storm and most were working-class. However, also included were a number of construction laborers who came to the area in search of work and, therefore, were not legally entitled to stay at a tent city. They usually presented themselves to the Red Cross, claiming that they had been living in the area, often providing a false address. When FEMA eventually took charge of admissions, they spent some time trying to weed out these and other "non-qualified" tent city residents.

While the Red Cross seemed to allocate individuals and families to the multi-family tents without regard to ethnicity, a pattern of segregation was discernable to the Campbell Drive tent city residents – especially the African-Americans and Haitians directly affected. When many houses in the primarily African-American area of Goulds were condemned at the same time that a number of undocumented Haitian workers began to arrive in the camps, both groups were assigned to a grouping of tents furthest from the camp entrance and the dining areas. This was interpreted by some as an intentional act of segregation and discrimination.

The Red Cross workers, and later the FEMA officials, in charge of the camp placements told us they occasionally received requests for tent transfers because occupants could not get along. The conflict that triggered these requests was often couched in ethnic terms. Complaints came from Latinos who did not want to share a tent with African-Americans; from African-Americans who did not want to share a tent with Latinos; from Anglos who did not want to share a tent with either; and so on. There was serious – if sometimes latent – ethnic hostility in the relations of tent city residents (Yelvington and Kerner 1993).

Researchers heard veiled references to "a certain kind of people," or "you know the kind I'm talking about," when the speaker was referring to another ethnic group, usually assumed to be from a lower socioeconomic class as well. Many of the tent city residents we interviewed complained of a "Latin takeover." This fits with the local discourse among non-Latinos, many of whom claim that Latinos have a disproportionate amount of local economic and political power. An interview with Wendy Johnson illustrates the Anglo/Latin tensions or, more precisely in this case, Anglo prejudices. This 33-year-old Anglo mother and her 4-year-old daughter were evicted from their Miami Beach apartment three weeks after the hurricane when she lost her job and could not pay rent. They were admitted to a tent city by the Red Cross and she later applied to FEMA for aid. According to her, the FEMA officials said she could be eligible for aid because she was a resident of a tent city. Although she was not a disaster victim per se, Johnson said she felt accepted by fellow Anglos in the tent city. She attributes this acceptance to the feeling among many Anglo tent city residents that Latinos – and especially undocumented workers – were being unfairly favored by the relief bureaucracy.

> I feel very fortunate that they did take me because . . . I do not bear a Homestead address nor was I a disaster victim, quote-unquote. But the people that I . . . associate with in the community – that I've met – don't mind my [being in the tent city] . . . there's a lot of, you know, "I'm not a disaster victim, so why am I reaping the benefits of a disaster victim?" And you know what they told me? I'm White [while] the majority of this camp – well, it's Mexican migrants. Uh, there's a real racial thing every-where around here. Like me, like Sue is American, a White American. She's a minority, she's from Homestead. Like I said [earlier], I know Homestead. I've been down here. But she really knows it. It's a very local,

small community. And she is a minority. She says we are a minority. They don't feel that we're being treated properly.

She went on to recount a story of a community meeting held the previous night. She had not attended, but told us what she heard from an Anglo friend who had attended. She claimed that issues were being discussed in Spanish and were not translated into English. She expressed her fears, and the fears of fellow Anglos, that they would be the victims of prejudicial treatment once the management of the tent city was turned over to the residents, the majority of whom were Latino.

> It's too Latin-dominated, but the majority always rules in a democratic society or whatever, and so these people really feel like they're gonna get the brunt of it . . . There's going to be jobs available here in tent city and they really don't think they're gonna get them. They really don't. Because they really feel that everyone is just going over and beyond themselves to accommodate illegal aliens, Mexicans, alright. Even at the community meeting last night they were saying, "Well can we just take our FEMA money and just go home?" . . . Now we have the illegal aliens that aren't supposed to be here and we're givin' 'em, we're setting them up totally. Then you have these other people that are involved in like Section 8 housing, HUD housing [the Department of Housing and Urban Development, i.e. public housing] and – I'm learning all this from the people that are involved in all these things – and they're mad because they even think the government is giving people too much.

This passage not only reveals views on ethnicity that Johnson shared with many other Anglos in the tent cities, but that she believes the Red Cross officials gave her a place in tent city because she is White. This selectivity on the part of Red Cross officials was not confirmed by our observations. In fact, many of the Red Cross workers we interviewed were at pains to ensure smooth ethnic relations and lamented the fact that there were instances of ethnic strife. While the interview with Wendy reveals her personal perceptions, a more active mode of prejudice surfaced in other interviews.

Billy Moss, a 36-year-old Anglo shrimper, told of his and his tent-mates' strategy for keeping non-Anglos out of their tent. There were only five people living there and when asked how they kept others out, he replied:

> We keep the extra bunks in to keep all the filth out, so to speak . . . I know that's bad to say, but its pretty much an all-American tent and we've kind of kept it that way. Yeah . . . there's not gonna be . . . as much noise. The language you can speak. There's not bickering here, bickering there, nitpicking and stuff like that. And that's the way we've kept it. Next door they've got this big old radio. They like to mess with us, turn it up real loud 'til 10 or 11 o'clock at night, and we'll throw cheese or something over there . . . you know, stuff like that.

On one occasion, while walking in an area assigned to families, one of our team members heard a song coming from the single men's side of the camp. The song was a rap-style song and the lyrics were very vulgar and specifically demeaning to Mexicans. The chorus went, "motherfucking Mexicans got everything, niggers ain't got shit." The tape was broadcast on a portable tape player very loudly from outside a tent, and was aimed in the direction of the family tents where many residents were Mexicans. It was not clear whether this tape was commercially or personally made. During an interview with a Puerto Rican family, we were told that the tape was regularly played loudly, day and night, and disturbed them greatly.

Frustration felt by the tent city residents did not always revolve around the issue of ethnicity. Many residents, especially mothers, felt frustration at what they saw as the degree of loss of control over their children. Several complained about the fact that their children were able to leave the tents through the side flaps and roam the camp freely. Indeed, large groups of unsupervised children were always roaming the public areas. After school resumed, these groups could be seen forming in the afternoons after their school buses dropped them off at the camp. Camp administrators saw these groups as potential problems and were at pains to find activities for the children.

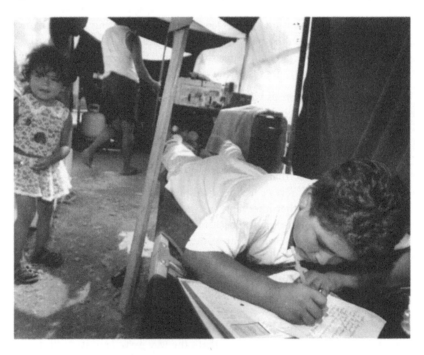

Plate 6.2 The routines of daily life, including school homework, continued for families living in the tent cities
Source: Hector Gabino/*Miami Herald*

104

Gloria Pagan, a 23-year-old single mother of four children, ranging in ages from 10 months to 7 years, expressed concern about her children. "All [the children] do is fight all day. They were always so calm and now they just fight all day . . . " [Interviewer: "How has it affected you personally? What are you feeling?"] "I feel mad." ["Who are you mad at?"] "At everything, they took everything I had. All I have now is my children . . . it's so hard, you know. I never thought this would happen to me. I am here with my mom and everything, but it's still hard when you got so many children." When asked whether the children's father had helped since the storm, she replied: "He came down to Homestead, but he wasn't with us. He does come around, but he hasn't been very helpful." ["Do you have any hopes for any help from him in the future?"] "Not really."

Other camp residents agreed that mothers with day-long primary childcare responsibilities had an especially challenging time. As one camp resident, Bud Talley, said:

> I notice that in the daytime, the women that have children here . . . that don't go to school, they are kind of stuck in a tent. A lot of the couples are starting to bicker and fight and you see the ladies howling at the children more often than they did the first few days. Things . . . are getting tense, and I contribute most of it to the heat . . . . The heads [toilets] are pretty nasty and they stay nasty although [the relief workers] do the best they can to clean them out. I can't imagine how they can let the kids go and feel safe about it."

One African-American mother found sitting with her infant child in an otherwise empty tent on a sweltering afternoon told an interviewer about having to wait for her other children to come home from school before she could go to the bathroom. She couldn't leave her child alone and she refused to take it into the dirty portable toilets. She refused the interviewer's offer to watch the baby for her for a few minutes, saying she was afraid an HRS worker might come by and accuse her of abandoning her child. Yet, this same woman remarked that she had warned her children not to get used to how good things were in the tent city. They would have to leave soon.

The Marines and Army personnel spent a lot of time with the children and many friendships developed. One of the most memorable photos to come out of Andrew was the one of a soldier holding a small child which appeared on the April 1993 cover of *National Geographic*. The military's efforts in this regard – certainly not officially part of their mission – were mentioned with appreciation by almost every resident we interviewed. We talked several times with a Marine Captain who was with the first unit to arrive and the last to leave. When asked who were the most appreciative residents, she replied:

> I think the families, that's the single group . . . the families. 'Cause the Marines really became part of their lives. We adopted families. We became

very involved with the kids and then through [them], their parents . . . you've taken this huge burden off their shoulders because you love to go out and play with their kids. And then, they don't worry about their kids.

We observed Marines playing ball with children, helping them with their home-work, carrying them on their shoulders, and stopping by tents to check on specific children to see if they got home from school alright. During the first few days when it became clear that they were going to have lots of bored children on their hands, the Marines built a playground. When children fell over the tent stakes, they covered each one with a plastic gallon milk container painted a bright color. The day the Marines left was marked by lots of tearful goodbyes.

One issue that seemed to occupy much camp discourse was food. Until the military established mobile kitchens and mess tents that made hot meals, residents ate packaged military food or MREs (Meals Ready to Eat). Initially, residents who had brought portable barbecue grills with them were allowed to cook for themselves, but after a fire occurred in one of the tents, this practice was banned. While many residents were indeed thankful for the military food, feelings of gratitude turned to criticism. As one man told us in Spanish, "We are not used to this kind of food. We are used to Latin food. You know, beans and rice. We are not used to this military food."

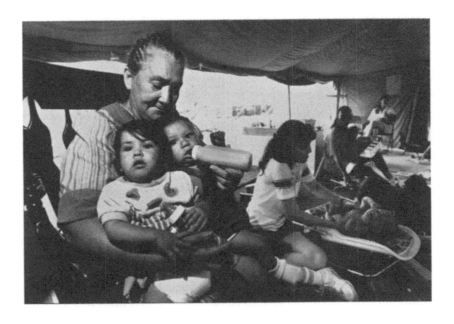

Plate 6.3 Caring for infants and small children at the tent city set up by the military at Campbell Middle School
Source: Chuck Fadely/*Miami Herald*

106

But beyond this were issues of privacy. One woman expressed what was one of the most trying and impersonal experiences: taking communal showers. She said, "So we are still . . . standing right in front of a woman taking a shower. Trying to shave or whatever a woman does . . . feminine hygiene and stuff." Communal living became very trying when there was a lack of respect for places where private daily functions had to be carried out in public. As one newspaper article reported about the Harris Field tent city:

The most hated man in the camp – nobody knows who he is – is the fellow who showers during the 6 a.m. cycle and moves his bowels at the same time, leaving the evidence for others to see and rejoice over when they arrive. "When I find him, I'm going to kill him," a resident was quoted as saying.

(Browning 1992: 27A)

Administrative mistakes sometimes contributed to problems of personal space. There was the case of a woman who was living in a tent designated for families with her husband and two young sons. Two other couples were allowed to move in. After several nights of the other four drinking alcohol and smoking marijuana, Marshals stepped in and confiscated the liquor and drugs. But, as the newspaper report stated:

One night later Candy and her family heard the unmistakable squeak of a cot moving up and down – eeaa, eeaa, like a metronome. Then the panting started. It drowned out the squeaking. "Two couples and an extra man were going at it," she said. "I couldn't believe it." The Marshals stepped in again, but said it was absolutely legal to have sex in a tent. It was happening discretely and not so discretely under tents all over the place. Kids were instructed to look the other way. You just can't ban sex if it's not lewd and lascivious and clearly not out in the open. Their hands were tied, they said. The next night one of the men slugged one of the women in the head. "That's when they came and took them out in hand-cuffs," Candy said. "But they were back three days later."

(Alvarez 1992b: 27)

DEALING WITH THE ISSUES

We observed numerous conscientious attempts on the part of camp officials to rectify problems brought to their attention. Military and camp officials held "town meetings" to discuss the changing situation with residents, to hear complaints, and to address problems. Officials announced their intentions to turn over a number of responsibilities for tent city life to the residents themselves. Committees of interested residents were formed to investigate and report on problem areas. In a town meeting we observed at the Campbell Drive camp, perhaps fifty people attended. The sides of the tent were rolled up and a number of folding chairs were placed in rows on the platform and on the grass surrounding the tent. There was a long table on one side where the officials sat. Business was conducted in English with a verbatim translation into Spanish by a volunteer interpreter. When meeting-goers spoke in Spanish, their remarks were usually translated for the officials and audience.

At this particular meeting, officials heard concerns about the dangerous crosswalk from the Campbell Drive camp to the school across the street and promised to station crossing guards there. They also heard complaints by women that male residents were leering at them when they walked to use the portable toilets at the edge of the camp. One woman suggested that toilets designated for women and for men be placed at opposite ends of the camp. She exclaimed in Spanish, "*¡De esa manera, podemos usar una silla seca!*" ("This way, we can use a dry seat!"). Another woman took her half-joking remark seriously. Before the first woman's remark could be translated and the laughter explained to non-Spanish speakers, she stood up and explained to the officials, "*Bueno, tú no tienes que sentarte completamente. Tú puedes simplemente agacharte ahí, como un helicóptero.*" ("Well, you don't have to sit down completely. You can just squat over there, like a helicopter").

CHANGING OF THE GUARD

About three weeks after the tent cities were established, the Red Cross volunteers in charge of admitting new tent city residents were replaced by FEMA. Rumors were that they accused the Red Cross of letting things get out of control. This policy shift had important implications for tent city residents. FEMA officials recruited a number of local Internal Revenue Service workers, mainly auditors, for intake responsibilities. They were given official leave from their jobs and paid overtime by FEMA. When we observed these workers, they were especially strict in terms of the identification and proof of pre-hurricane residency requirements. Moreover, it was determined that applicants had to be found eligible for FEMA aid before they were admitted into a tent city, which effectively precluded some of the most needy from having access to tent city shelter. The function of the tent cities went from a caregiving one under the Red Cross to an eligibility-establishing one under FEMA.

Even when FEMA provided aid in the form of a reimbursement check, their regulations, at least initially, showed a remarkable lack of fit with local conditions. While aid was given primarily to homeowners, some was also available for loss of household property, such as furniture. Of course, there was no place to buy furniture and, for many, no place to put it. Under FEMA regulations, reimbursement checks were made out to the designated head of household (typically the first person from an address to register) and a household was defined to mean a physical unit or dwelling. When two or more families shared a single apartment, yet owned furniture individually, as was the case among many of Homestead's migrant workers, only one check was made out. Tent city residents told many stories of persons claiming to be the head of household and then absconding with the reimbursement check intended for all of the apartment's dwellers.

Some tent city residents noted the FEMA take-over by pointing to what seemed to be a more strict rationing of supplies. Others complained of FEMA policies designed to restore the *status quo ante*, rather than assist more needy victims. When telling of her encounter with FEMA officials, tent city resident Vickie Lloyd said,

> I'm not a very patient person. I get aggravated very fast with this bureaucratic bullshit, excuse my language, but that's how I feel about it. And we're still here . . . They're bringing mobile homes in here. I don't know what they're gonna do with them. They are offering mobile homes, if you have a place to put it. So, the kind of impression I'm getting is this whole thing is sort of for the homeowners, those that own their own houses, mobile homes, whatever. If you didn't, you're kind of shit-out-of-luck . . . 'cause all these homes here are for people that have land. If you don't have a place to put it, you are still in the tent city.

Ironically, we heard of complaints among victims still living in their damaged houses that tent city dwellers had an unfair advantage and received FEMA trailers more quickly. In media reports, FEMA officials said that, while it may have appeared that tent city residents were among the first to receive trailers, in fact FEMA had been assigning trailers on a first-come, first-served basis and the tent city dwellers were simply among the first to apply. However, a FEMA Recovery After Action Report contradicts this:

> Unfortunately, because of political pressure to close "tent" cities, those occupants were given first priority for placement into available units. Although "tent" city occupants had safe, sanitary, emergency housing, albeit tents, the majority of the rest of the disaster victims were living in their damaged dwellings with no resources.
>
> (Polny 1993: 56)

Advocacy groups complained that the biggest obstacle was the FEMA rule that stated only heads of household – as defined by FEMA – were eligible to receive trailers and other forms of aid. María Escobar, with the South Dade Immigration Association, said, "The reality is that the households are not Mom and Dad and three kids. These people rent rooms and part of rooms. They don't have rent receipts" (Alvarez and Higham 1992: 6A). In households shared by three or four families, FEMA either denied the entire household money or awarded assistance to only one family. A FEMA spokesman was quoted: "We've encountered some unusual living arrangements. We're being as liberal as possible. We walk a fine line between protecting the taxpayers from fraud and making sure people entitled to benefits get them." (Alvarez and Higham 1992: 6A). As discussed throughout our work, given the complex ethnic, cultural, and class makeup of South Dade, official policies often did not match the realities of victims' lives.

THE TENT CITIES CLOSE

The tent city population began to dwindle by the end of September due to FEMA restrictions and the fact that many residents were able to make other arrangements. Many utilized aid money from FEMA and the Red Cross, as well as food stamps and other kinds of public assistance, to relocate. FEMA transferred some families into FEMA-subsidized mobile homes, either in parks in South Dade or at the site of their damaged homes. A year after Hurricane Andrew there were 2,261 FEMA trailers throughout South Dade, but this in no way met the demand (Browning 1993).

By the time the tent cities abruptly closed in the third week of October 1992 – just two months after the hurricane – only the poorest of the poor and those most devastated by the storm remained, though there were many, many more needing continued assistance. We were there the last day, 23 October, when the very last residents, Victor and Sonia Welsh and their six children, loaded their belongings into a FEMA van and headed to the New Life Family Shelter in downtown Miami. They hoped the move would help them find a permanent home and employment for Victor (Hartman 1992).

A minority of community voices expressed concern for what would happen to families and individuals living at the tent cities once they were dismantled. There was no available housing for them in South Dade. In media reports, FEMA officials claimed that almost all the people who stayed in the tent cities received FEMA trailers and other assistance upon the dissolution of the temporary shelters. Some homeless advocates said that the federal government closed the tent cities 11 days before the presidential elections to give the impression of federal effectiveness, though this was denied by a Bush administration spokesman.

When post-storm efforts in the private and public sectors turned to the housing problem, they sought to provide shelter – some of it permanent – for the thousands of migrant laborers who lived in the area. The large agricultural businesses in South Dade, dependent on migrant labor, lobbied for the county to build permanent housing to replace the mobile homes destroyed in the storm. As the tent cities were closing, more than 7,000 agricultural workers were expected for the winter crop (*Miami Herald* 1992m, 25 November). With federal and state money, the county moved quickly to create a mobile home village and to solicit bids to build permanent migrant worker housing. In effect, the sheltering of these workers subsidized agribusiness (see Brennan 1992a; Brennan 1992b).

The closing of the military tent cities was immediately followed by official requests for a new tent city since it was estimated that 2,000 people were still living in their cars or in makeshift camps in South Dade. The Dade County Director of Homeless Programs asked FEMA to re-open an emergency tent city for the homeless in mid-November. FEMA refused, saying the need was not there. Miami had always had a relatively high homeless population, partially because it is a destination for migratory homeless. A 1990 countywide survey had found 5,000 homeless; a 1991 survey counted 6,000. The hurricane certainly added to this (Merzer and Viglucci 1992). In early November, a federal judge ruled that Miami's homeless had a constitutional right to reside on public property and ordered the City of Miami to provide two safe havens for their use: a park and under an expressway overpass. It was not clear how this ruling would affect those in South Dade made homeless by Hurricane Andrew. According to a FEMA spokesman, "We don't see a need for another tent city for the disaster homeless. Tents are not a real good solution" (Alvarez 1992c: 6A). The county then asked FEMA to turn the tents and equipment over to the State of Florida so the county could set up another tent city.

There were widespread calls from community activists and progressive civic leaders to establish new tent cities as soon as the FEMA tent cities were dismantled. Representatives of the Archdiocese of Miami's Catholic Community Services reported that its five missions in South Dade had been besieged by those forced to leave the tent cities, but still in need of shelter and assistance. In conjunction with Metro Dade County and the Florida Department of Health and Rehabilitative Services, the Archdiocese conducted a survey of the homeless in South Dade in December and counted more than 5,000, approximately 2,000 of whom were agricultural workers, 2,000 were out-of-town construction workers, and 1,000 were Dade residents dislocated by the storm. Most of the latter were intact families. The county sought and received $1 million from the federal government to establish a new tent city. The plan was to maintain a tent city of 100 families for six months, operated jointly by the Archdiocese and the county, providing a range of social services to foster independence. However, the Life and Family Support Center lasted only seven weeks and served only 52 families – a victim of more bad weather and more bureaucratic wrangling (see Semple 1993).

Nearly all of the federal and state public housing units in South Dade were badly damaged or destroyed. Many families had no choice but to remain in damaged units for months, often without electrical power or telephone service. For example, in December it was reported that more than 300 people were still living in county-owned homes that Metro Dade acknowledged should be condemned. "In some places, [people] live nine to a room. In others, rats and cockroaches scamper through three-foot-wide openings in walls. Smashed cars, trashed tricycles and tons of garbage fill their yards" (Swarns 1992: 2B). While some public housing recipients had been given grants from county and federal sources to acquire new living quarters, most were forced to stay in severely damaged residences. Many whose apartments were condemned became home-less. By December, the government had offered more than 900 rent vouchers, but only 35 had been used. The reasons for this low-level utilization are numerous. Residents reported that landlords would not accept their federal rent vouchers. Others stated that they did not want to move away from their neighborhood, and, presumably their support networks (Strouse 1992). A third problem with the voucher system was that while the federal government promised 13,000 vouchers that would pay up to 70 per cent of a storm victim's rent for two years, the vouchers were still stuck in Washington two months after the storm.

A fourth problem of the voucher system was simply the lack of housing (see Swarns 1992). By December 1992, the occupancy rate in the Dade County apartment market approached 100 per cent. The Greater Miami Apartment Association, which had a list of over 6,000 landlords, put the availability rate at 0.01 per cent. Two research groups put Broward County's post-hurricane vacancy rate at about 1.6 per cent, compared to the reported rates of the previous November of 5.5 per cent for both Dade and Broward. An owner of a rental property management firm estimated that overall rental prices had risen by 15 to 20 per cent since the storm (Hartman and Penn 1992). In South Dade, landlords continued receiving insurance and assistance checks and were eager to start repairs. They began evicting tenants so that repairs could be made, creating more homeless families. While this was a continuation of the process that sent families to the tent cities, this time they were not available.

CONCLUSIONS

The tent cities case supports the theoretical proposition that a conflict – or at least a non-convergence – of interests characterizes a population both before and after a natural disaster. In the tent cities, overt and covert conflicts were apparent at many levels, even if temporarily suspended by a sense of "communitas" (Turner 1969) among the Hurricane Andrew's victims. The victims' social and cultural position prior to the storm appeared to dictate not only who would end up in the tent cities but, in some cases at least, the perception of who was relegated to the least-desirable sections within the tent cities.

The preexistence of ethnic competition and conflict is evident in the socio-logical, anthropological, and even literary work on Miami (for example Grenier and Stepick 1992; Portes and Stepick 1993; Didion 1987; Allman 1987; Rieff 1987, among others). These rich studies add much to our knowledge on ethnicity and ethnic conflict in Miami – known variously as a "multi-cultural brew"; a "salad bowl" (where the ingredients mix but remain distinguishable); or an "ethnic cauldron on the point of boiling over." But there is a danger in interpreting these findings as preponderantly horizontal differences among ethnic groups.

The identification of a certain region with certain cultural features is a common tendency in anthropology. Arjun Appadurai has criticized the anthro-pological tendency to emphasize certain cultural features of a geographical place so that the place comes to stand for these features within the discourse of social scientists. In this tendency, a few simple theoretical handles become metonyms and surrogates for the civilization or society as a whole: hierarchy in India, honor-and-shame in the circum-Mediterranean, filial piety in China are all examples of what one might call gatekeeping concepts in anthropological theory, concepts, that is, that seem to limit anthropological theorizing about the place in question, and that define the quintessential and dominant questions of interest in the region.

(Appadurai 1986: 357)

For Appadurai, this tendency has two implications:

One is that the discussion of theoretical issues tends (surreptitiously) to take on a restrictive local cast, while on the other hand the study of other issues in the place in question is retarded. Thus the over-all nature of the anthropological interpretation of the particular society runs the risk of serious distortion.

(Appadurai 1986: 358; see also, Fardon 1990)

In the study of Miami, with the general emphasis on ethnicity at the expense of an approach that incorporates political economy, ethnicity becomes one such gatekeeping concept. I do not deny the continued salience of ethnicity in South Florida. Indeed, our research shows that issues of ethnicity were extremely important to tent city residents. Yet, this conflict must be placed in a larger class context, taking into account the role of the state as a critical actor. It is hoped that this case study opens up fields for further investigation into the political economy of the area.

What emerges from our research is a picture of a region not only culturally pluralistic, but encompassing serious inequality. This is best explained as a "culture of ethnicity," where certain ethnic identities and ways of life are given more value than others and simultaneously, where cross-cutting alliances of varying duration between individuals, sometimes groups, exist to, in a way, mitigate the effects of widespread relations of dominance. This ethnographic study suggests the kinds of forces that conspire "offstage" to impact what ethnographers see "onstage." The paramount role of the state in this drama, in the form of disaster relief agencies and their policies, is obvious. It is evident, for example, that the policies of FEMA and other official agencies not only reflected some misguided notion of the integrated, functional community, but that this fictitious "community" conformed to an idealized North American middle class model of household and family. If the model upon which such assumptions are based does not even accurately reflect the modern realities of the middle class, how much less so can it be applicable for the people on a country's cultural and economic periphery? While the tent cities provided temporary relief for the poor and powerless, the policies that determined admittance to the tent cities served to exacerbate inequality. This was because the policies were based on eligibility criteria for an official relief package that was founded on erroneous notions of community.

113

The analysis of state policies must be an integral component of scholarly research on disasters and on South Florida's cultural and ethnic diversity. There is a joke told in South Florida that on returning to his Latin America homeland, a visitor tells his friends he liked Miami because it has good restaurants, good shopping, and besides, it is so close to the United States! Our work suggests that no matter how unique visitors, residents, and researchers find South Florida, it is part of a circuit of social relations and cultural arrangements that reach far beyond the boundaries of Dade County, articulating with the rest of the United States. In this study of the nature of United States disaster relief policy and practice, we can see how specific FEMA policies are not designed to improve the status of the needy, but to return to some preconceived prior state of normalcy that was actually nonexistent.

In making applied recommendations, there is a tendency to become so involved in empirical details that we lose sight of the broader implications of a policy for those it is directed toward, as well as the implicit theories, such as the notions of family and community, behind the said policy. Anthropology holds that researchers should incorporate the viewpoints of those being studied when searching for applied solutions. In making brief policy recommendations here, I have incorporated issues of importance to the tent city residents themselves, as well as issues that arose in the experiences of the fieldwork team. As such, perhaps both visions are incomplete, but they are at least complementary and go beyond the operating assumptions that were in place at the time of the hurricane.

SOME POLICY CONSIDERATIONS

- Social service providers and disaster relief professionals need to be knowledgeable about the linguistic complexity of the United States and to have multilingual dissemination of information from the onset.
- While housing assignment of individuals and families to tents should not be made on the basis of ethnicity, officials need to be aware of possible ethnic conflict and to take measures designed to alleviate it as much as possible. On the other hand, the "ghettoization" of groups assigned to particular areas of temporary shelters should be avoided since this encourages not only polarization, but may lead to official ethnic discrimination. In both instances, officials need to be perceived as impartial and aware of the special needs of all groups.
- Officials should consider the special needs of specific populations, including the elderly, the handicapped, and children, when planning temporary shelter.
- Officials should not operate from the assumption that all households conform to a middle class model of the nuclear family. Further, they should be aware that within the same household all members' interests, wants, and needs are not necessarily the same. This is especially true for issues relating to gender and the relations of power between household members.

114

• Finally, and most importantly, there needs to be an expanded definition of need on the part of disaster relief officials. The research shows that there was a hierarchy of relief, with the most needy sometimes going without aid because they did not meet the narrow eligibility requirements set by FEMA and other organizations. The existence of more than 5,000 homeless, many of whom were storm victims, on the still rubble-strewn streets of South Florida more than a year after the tent cities closed down is evidence of this policy of negligent exclusion of the poor and most powerless.

NOTES

1 The tent city interviews and ethnographic observations were conducted by Manny Alba, Linda Beer, Chris Girard, Donna Kerner, Barry Levine, Betty Morrow, Walter Peacock, Jim Riach, and myself during September and October 1992. To protect the anonymity of our respondents, pseudonyms are used. I would like to thank Neill Goslin for his excellent editorial assistance in preparing this chapter.

7

A GENDERED PERSPECTIVE

The voices of women

Elaine Enarson and Betty Hearn Morrow

Social structures not only provide the context, form and meaning for response, but are a critical part of vulnerability. The vulnerability of women stems from cultural, political, and economic conditions. The poor and destitute are the most vulnerable, and they are disproportionately women and their dependent children.

Document of the UN Disaster Management Training Programme
(Wiest, Mocellin and Motsisi 1994: 11)

The "official story" of disasters generally overlooks women's experiences as victims, as well as responders. In this chapter, we use the case of Hurricane Andrew to illustrate the need to account more effectively for the experiences and insights of women if we are to understand better households and communities hit by disaster. Our qualitative findings are synthesized around four composite profiles which we develop to introduce some important themes which emerged from our work. In keeping with our goal of illustrating the potential of a gendered analysis for advancing theoretical and empirical work, as well as our argument that a better understanding of women's diverse experiences will promote more effective disaster planning and response, we end by suggesting a series of research questions and proposing a set of policy recommendations to disaster planners and responders.

CONTRIBUTIONS OF A GENDERED ANALYSIS

Most disaster work assumes a gender-neutral social system in spite of a growing body of research documenting the significance of gender and gender relations in social life (Hess and Ferree 1987; Epstein 1988; Anderson 1988). The complexities of social structure and culture typically result in different social environments and world views for men and women. It follows, then, that there exists a gendered dimension to the effects and responses associated with any social event, particularly one as significant as a disaster (Morrow and Enarson 1994). As Shaw (1989: 13) states, "In any society in which elaborate gender domains are constructed, then both hazards and relief measures will be 'gendered' with different consequences for men and women." This implies more than a simplistic bipolar view of gender as might be evidenced, for example, by simply comparing survey responses of male and female victims. Contemporary gender studies, while emphasizing the unique experiences of women and men, also demand an exploration of the complex intersections of gender with other social dimensions such as race/ethnicity, culture, and class (Brydon and Chant 1989; Ward 1990; Collins 1990; Peterson and Runyan 1993; Zinn and Dill 1994).

To understand fully household disaster response, an accurate analysis of patterns of domestic labor and decision making is essential. Without generalizing about women as a class, several basic social arrangements persist across a wide diversity of cultures. As mothers, partners, daughters, grandmothers, sisters, and aunts, women continue to provide the bulk of household labor and family caregiving (Finch and Groves 1983; Abel and Nelson 1990), regardless of their participation in the formal labor force (Hochschild 1989; Shelton 1992). Additionally, the proportion of households headed by women has increased dramatically – it is now 25 per cent of all US households, even higher among minorities, the elderly, and the poor (Scott 1984; Sidel 1987; Ahlburg and DeVita 1992; O'Hare 1992). Outside the household most caregiving work is also done by women, although this is not reflected in the lines of control and authority within caregiving occupations and organizations (Acker 1991; Reskin and Padavic 1994). While playing crucial public and private roles, women's voices have been largely absent or ignored in organizational and community policy-making, including decisions about disaster response and recovery.

THE NEGLECT OF GENDER IN DISASTER RESEARCH

The effects of gender and gender relations have been virtually ignored in most disaster research, with few sources addressing women's wide range of involvement in disaster-stricken households and communities. Women and gender still remain largely absent even as organizing categories in the disaster literature. While sex as a bipolar variable is sometimes analyzed in quantitative studies, a complex gendered analysis is rare. Attempts to quantify household recovery by measuring domestic assets, for example, usually neglect the human resource of domestic labor, such as how much time women have available for household recovery activities (Bates and Peacock 1993). In contrast, the dimensions of race and ethnicity, culture, and social class are increasingly recognized as significant factors influencing household and community experience and recovery (Drabek and Key 1982; Peacock and Bates 1982; Bolin and Bolton 1986; Perry and Mushkatel 1986; Oliver-Smith 1990; Perry and Lindell 1991; Peacock, Gladwin and Girard 1993; Phillips 1993b; Blaikie *et al.* 1994).

There are exceptions to the neglect of women in disaster research. (For a literature review on gender and disaster, see Fothergill 1996.) In his classic analysis of the Buffalo Creek flood, Erikson (1976) relates aspects of male gender identity to psychological disaster recovery. Poniatowska's testimonial from the 1985 Mexico City earthquake is rich with the experiences of women, as both victims and rescuers (Poniatowska 1995). A video documentary by Carol Ward poignantly captures the experiences and emotions of women from a small fishing village in South Carolina after Hurricane Hugo (Ward 1990). Gender differences in the consequences of disasters and various assistance programs are beginning to be examined (Schroeder 1987; Shaw 1989; Phifer 1990; Nigg and Tierney 1990; Chowdhury et al. 1993; Khondker 1996). In the area of family response, several studies suggest that crisis reactions tend to follow traditional gender roles which, in turn, limit their effectiveness (Hill and Hansen 1962; Drabek 1986; Hoffman 1993). Analyses of household evacuation decisions suggests that women consult more frequently with relatives and friends and are more apt to believe and heed a warning (Drabek and Boggs 1968; Turner et al. 1981; Neal, Perry and Hawkins 1982).

The high degree to which women are active in community organizations associated with disaster issues is beginning to be recognized (Neal and Phillips 1990; Leavitt 1992) and was especially well documented after the 1985 Mexico City earthquake (deBarbieri and Guzman 1986; Rabell and Teran 1986; Massolo and Schteingart 1987). Furthermore, international relief agencies are increasingly recognizing the status and role-related vulnerabilities faced by women during emergencies, as well as their special needs as refugees (League of Red Cross and Red Crescent Societies 1991). It is also important that women's capabilities and informal mobilization efforts be effectively utilized in disaster response (Wiest, Mocellin and Motsisi 1994).

Important as these exceptions are, it is hard to escape the conclusion that theory and research in disaster studies have generally failed to acknowledge the myriad, but not always obvious, ways in which gender impacts the lives of victims and responders. This neglect, along with the gender bias sometimes attached to disaster research, results in programs which do not adequately reflect an understanding of female victims' daily lives and which fail to utilize effectively the knowledge and skills of over half of the population.

A QUALITATIVE STUDY

In order to understand better the implications of gender and the roles of women in household and community preparation, relief, and recovery efforts, we conducted a qualitative sociological analysis of women's experiences in the aftermath of Hurricane Andrew. Data were collected through interviews and focus groups with victims and service providers, observations in the tent cities, service centers, provider organizations, and at meetings of emergent community groups. We also drew from other projects of the FIU Disaster Research Team, including over fifty interviews in the tent cities and more than forty interviews with agency caseworkers. Our goal was not to represent women as a group or to speak for most women, but rather to study selected segments of the victim and caregiver populations whose circumstances and experiences provide important insights and perspectives. We interviewed immigrant and migrant women from Haiti, Cuba, Mexico, and Central America, African-American single mothers and grandmothers, women construction workers, business owners, agricultural workers, teachers, social workers, battered and homeless women.

We have organized our findings around the experiences of four fictionalized women – a social worker, a single grandmother in public housing, a relocated trailer camp resident, and a small business owner. In a field often dominated by the abstract and impersonal, we present the particularistic first-person voice by introducing the many women we met as composites, rather than as literal portraits. This technique protects individual anonymity while personalizing the themes addressed. The portraits are representative of the situations and experiences of women we interviewed, heard about, read about, and observed during a period of great crisis. We weave and merge observations and feelings selectively, carefully reflecting the diversity of our sample and avoiding unrepresentative implications. We focus on patterns of common situation and experience to theorize about women's disaster experiences and, more broadly, about how gender, race, and class relations interact in disaster-impacted communities and households. All quotations are transcribed from recorded interviews and are repeated verbatim.

THEY SPEAK FOR MANY

Irene Phillips: social worker

A year and a half after Hurricane Andrew, the couches and chairs are still covered with plastic amid great piles of building materials and tools dominating Irene's living room. Her husband works for a building company and has been repairing their house in his spare time, of which he has little these days. John was transferred north after his work site was destroyed, increasing his commute to over two hours each day. As a result, he reluctantly agreed to hire out some of the work, if skilled and reliable people can be found. This Anglo couple, in their mid-thirties, are long-time residents of a working class neighborhood of small homes in Cutler Ridge. They are the parents of a 16-year-old son and a 12-year-old daughter.

Sitting in the middle of construction materials and boxes, Irene tells us how lucky she feels – lucky to have been able to evacuate her family to her brother's home in North Miami; lucky to have had sufficient insurance to replace the roof and rebuild the four interior rooms; lucky to have a home that is still habitable. Before evacuating, her husband and son put up homemade plywood hurricane shutters and packed up the family's business and personal papers, thereby lessening their losses.

Irene was indeed luckier than those who had no help preparing for the storm. We learned of the futile efforts of a group of women in public housing to cover their apartment windows with leftover construction materials and of an elderly woman who tried to show her young grandchildren how to nail closet doors over their windows. In general, women living without partners are less likely to have the resources – money, transportation, and labor – to complete disaster preparations. Thus, their unprotected homes tend to sustain high levels of damage. Single women, particularly widows and mothers of young children, are especially vulnerable if they lack nearby kin. We heard repeated reports, one from a witnessing construction worker, of unscrupulous contractors systematically targeting single women in desperate need of home repairs. Women who spoke little or no English were particularly vulnerable to coercive practices. Contracts were signed in ignorance or under duress, only to have their hopes dashed when work was not completed and advance deposits lost when contractors skipped town. We found no evidence that authorities considered that gender might be a factor in contractor fraud.

Irene speaks in the voice of the "guilty survivor" – reluctant to complain about her living conditions when so many have it much worse. When not at her job, she works on the house with her son and husband and negotiates with suppliers and workers. She enjoys the hands-on work and speaks with a certain pride about managing the project, but it is hard and frustrating work, consuming every spare moment. "We do our planning in the mornings, 6 o'clock in the morning. We go through all the different things that have to be done for the day. In the evening when we get home we check the work that was done."

Their modest house has become cramped since the storm. When county inspectors declared that the nursing facility where Irene's mother-in-law lived was unsafe, their dining room was converted into a bedroom for her. While John does not help with her special care, Irene counts herself lucky that only one person came to live with them. As a result of the storm, six extra people have been living next door for over a year. Once happily past the labor-intensive years of having a young family, Irene's neighbor is again hard at work keeping a large household going, only this time under very trying circumstances.

In a street still only half occupied, the new household members next door provide playmates for Irene's daughter, but the child misses her old friends who moved away after the storm. She complains of headaches and sleeps a lot, but Irene expresses more concern for her older son, a high school senior. He hates the long bus trip to the undamaged, but crowded high school he was transferred to and constantly threatens to drop out to work on the house.

Our respondents cited numerous instances of school-age children suffering long-term effects from Andrew and of the tremendous strain felt by their teachers, counselors, and parents. The public school system reopened only two weeks after the hurricane destroyed or severely damaged over thirty schools, as well as the homes of many public school employees. (For an analysis of the school system's response to Hurricane Andrew, see Provenzo, Jr. and Fradd 1995.) Tens of thousands of children were uprooted, bused long distances, and then relocated again when old schools reopened. Younger children dissolved into tears when winds rattled the school windows. One teacher told of a first-grade student who, when given a donated set of crayons, quickly put them away in his desk. Asked why he wasn't using them, he replied, "I don't want them to blow away."

Older children felt the loss of peers, school activities, and recreational facilities. High school seniors graduated from strange schools where they didn't feel they belonged (Marks 1993). Homestead Senior High students spent an entire academic year without a school or community library. It was over a year before the first movie theater reopened in South Dade. The despondency and uncertainty experienced by young people was expressed in many ways, including depression, withdrawal, disruptive behavior, and violence.

Because authorities felt it was in the best interests of the children to begin the school year as soon as possible, teachers and staff were forced to teach under terrible conditions. We heard countless stories of their struggles to reach out to suffering students while dealing with major disruptions in their own lives, including distraught families, damaged homes, and lengthy commutes in impossible traffic. Their heroic efforts are well represented in this letter we received from a woman whose nearly destroyed home served as a refuge for an extended family of eleven people:

> After two weeks of exhaustive cleaning and primitive survival, some sort of normalcy returned to my life: I reported to work. The tired faces of our faculty reported to the puddle-filled, smelly [school] cafeteria. If those first two weeks of the aftermath had been difficult, it was nothing like the months to follow. As a hurricane victim I faced my own tremendous loss and frustrating aftermath. And as a teacher I had to leave my personal sorrows behind and face my classes of tired, mournful and bewildered student-hurricane-victims . . . Perhaps it was seeing my students' tremendous difficulties that gave me the strength to get up every morning, take a cold shower by candlelight, and report to work.
>
> (Colina 1995)

In the spring she was chosen Teacher of the Year.

Within days of Andrew, private and public agencies "set up their tents" and thousands of volunteer and paid workers attempted the daunting task of bringing services into the area. In Irene's case, only days after the hurricane she was back at work at the non-profit family service agency where she has worked for 10 years. Though it has been a full year and a half since the storm, Irene still puts in long hours dealing with an expanded case load. She confided about her struggle to fight off overwhelming feelings of helplessness and depression.

As caseworkers, nurses, counselors, and relief workers, women were instrumental in providing services. The concentration of women in human service occupations puts them at special risk during disaster recovery. The physical and emotional demands of the "double day" at work and at home expanded exponentially after Hurricane Andrew. Most South Florida organizations did not

Plate 7.1 Schools were opened only three weeks after the storm, often under very
difficult conditions
Source: Paul Rubiera/*Miami Herald*

anticipate that, as the recovery period lengthened, workers' emotional needs
would increasingly impact organizational effectiveness. While some agencies
eventually received funds for employee counseling and stress workshops, these
initiatives would have been more beneficial had they been available earlier.

Following a major disaster communities turn in many directions for help – to
government and their elected leaders, to state and federal agencies, and to the
military – historically male-dominated institutions. And yet, female employees
are a majority in many, if not most, organizations providing relief and recovery
services. While disaster work typically assumes a male persona in public
discourse, the skills and training, as well as social and emotional resources, of
women are central to both short- and long-term community recovery.
Nevertheless, with some notable exceptions, such as Beth Von Werne at Catholic
Community Services and Mary Louise Cole, Director of ICARE (an interfaith
coalition of groups involved in rebuilding homes), women were severely under-
represented among those making important decisions about community
recovery and rebuilding after Hurricane Andrew.

Disaster-hit communities can anticipate the loss of an often invisible
and always undervalued resource – the many volunteer hours women give to
community causes which so often supplement or replace direct social and community services. After Hurricane Andrew, schools, libraries, scouting programs,
churches, and health and social services agencies, such as the American
Red Cross, felt the absence of women whose volunteer labor was suddenly

redirected to meeting more pressing household needs. The influx of relief workers into a community is no substitute for the long-term loss of local volunteers. Disaster planners need to anticipate the loss of local women's voluntary labor after a major disaster.

Hurricane Andrew disrupted, but did not destroy, the routines of social life. Irene still treasures her Wednesday evenings singing with the women's choir which reassembled a month after the storm carried off the church roof. This group has always been an important support group for her, but never more so than now. Similarly, women throughout the area came together, sometimes for the first time, to respond to community needs. Haitian women from Homestead and Florida City organized a community festival celebrating Haitian culture and the coming of spring. Others designed and coordinated projects to replant public spaces, to clear playgrounds, to plan social activities – networking to rebuild a sense of community and regain perspective on their own lives.

Like so many women, Irene wishes she and her husband talked less about drywall and more about what they are going through, but she says John is even more reluctant now to talk about his feelings or to ask for help. Many observers noted that men seemed to focus on the instrumental tasks of rebuilding, both at home and at work, and to withdraw from their partners and children. The reported accounts of increased incidence of male suicide, alcoholism, and violence are vivid indicators of men's pain and distress (for example, Wallace 1993; Laudisio 1993; Strouse 1995). Future disaster research is needed to document the emotional responses of men, as well as the resources that sustain them. Speaking of John reminds Irene suddenly of the time and she apologetically ends the conversation. As we head toward the front door, she stops to proudly show us the new tile in the bathroom.

Pat Higgins: head of a multigenerational family

When the Army trucks finally rolled into the grounds near her public housing complex in an isolated neighborhood in rural South Dade, Pat and the other women of Garden Grove were ready. Days before they had organized a cleaning brigade and set up a mass kitchen for their housing project in the cafeteria of the badly damaged elementary school across the street. The troops were a welcome sight after days of struggling without outside help. Since their housing complex stood directly in Andrew's path, the experts estimated that it had been blasted with 175 mph winds (Sheets 1993a; Wakimoto and Black 1994). While thankful to have survived that horrific night and to still have a roof, albeit damaged, Pat, her two adult children, and three grandchildren lost nearly everything they owned. Clothing and possessions not destroyed during the storm were soon lost to the rain and mildew.

The women of Garden Grove, most of whom are elderly and/or have small dependent children, received no help in preparing their buildings for Andrew. The walls of their buildings had come through intact, but the unprotected windows were blown out early in the storm, allowing the wind to tear through the apartments, lifting the roofs off some. Shutters might have saved their homes. Ironically, the women knew that sheets of plywood were in a locked storage room on the property, but had no way to get it out to cover their windows. However, several days later Pat joined with other residents to break open the storage room door and use the plywood to try to close out the elements and secure their apartments. Pat is still angry with the property managers. "Everything was right there. All they had to do was open it up and give us some nails – we could have did it ourselves. They didn't want to do it. They didn't tell us anything."

Without electricity, telephones, or newspapers, the women relied on word of mouth to locate emergency and relief aid. With weary kids in tow in the stifling heat of late summer – when every tree had lost its leaves and the foul odors of decaying refuse thickened the air – they struggled to find drinking water, food, diapers, shoes, towels, medicines, and glasses. The Red Cross vouchers she eventually received didn't go far. "I had gotten vouchers – one for food. The food I couldn't spend because they gave it for a store that wasn't open . . . One for clothes, and one for linens. But all I could do was buy me two things and my son two things – him a sheet set for his bed, one for my bed, and four towels."

During the two months the military camped in Garden Grove, they provided a bright spot in an otherwise dismal world. The soldiers helped clear debris and secure their apartments, provided food, water, and first-aid. Perhaps even more important, they played with the children, listened to the adults, and provided hope. When they pulled out, however, little permanent progress had been made toward repairing their homes and community. Pat felt deserted and cheated, "I really didn't feel like – that we got everything that we should have got. But I know that they're trying to get us to be self-sufficient and start getting back on our feet . . . I think they left us too early." As an African-American head of household supporting a multigenerational family, Pat knows plenty about self-sufficiency, advocacy, and community organizing. Laughing aloud, she recalls the day she publicly rebuked government officials touring the disaster zone. Her advocacy efforts got results, "I went to the City Hall and had them take buses and vans to take the people to wash their clothes, take them shopping, spend their vouchers they got from the Red Cross and stuff like that."

Because Hurricane Andrew's victims were spread over such a large area, it was often necessary to travel long distances under difficult circumstances to reach disaster relief centers. Confusion about available relief and how to get it added to victims' frustration. Pat told us about the terrible day she took three different buses to reach a particular relief center downtown, carrying her infant grandchild, only to learn that she still was not at the right place. Once there, immediate help was rarely available – intake workers usually scheduled future appointments for home inspections or return visits to complete more paperwork. For months, exhausted relief workers labored long hours while exhausted victims waited outside in slow-moving lines. The atmosphere deteriorated as waiting periods

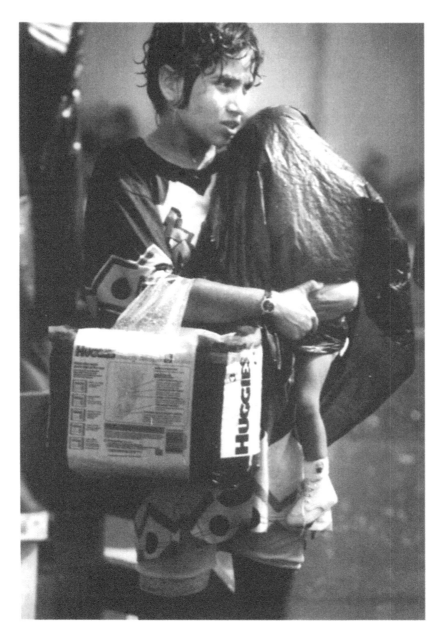

Plate 7.2 Waiting in the long lines, often in the rain, in order to receive supplies and apply for assistance was especially difficult for those with small children
Source: Charles Trainor, Jr./*Miami Herald*

lengthened and assistance declined. While the ill, heavily pregnant, or very frail were often invited inside to wait in air-conditioned comfort, most centers had no provision to help clients' children endure the long waits.

Interviews with Andrew's victims and workers confirmed our on-site observations that the tension level was high at many relief centers, often reflecting cross-cutting patterns of race, class, and gender. Relief workers, primarily White and middle class, typically came to South Florida from distant regions and had little, if any, experience working with culturally diverse clients. Many had been called from unrelated jobs – such as the Internal Revenue agents who became FEMA intake workers – and lacked any social services training. While many relief workers were patient and understanding, others were quick to judge poor and/or ethnic minorities as less deserving than victims who looked and acted more like themselves. We overheard workers' derogatory remarks about poor women's lack of English skills, large families, partners, or personal appearance – prejudices that negatively impacted service delivery. As one African-American woman said, "They hear you, but they don't *hear* you."

One legal aid worker noted that recent immigrants from countries with a legacy of political repression, such as Guatemala and El Salvador, were reluctant to seek help. Undocumented migrants were intimidated when uniformed immigration officers were deployed to distribute water and food immediately after the storm. Despite the large numbers of Spanish and Creole-speaking victims, translators were in short supply during the first few weeks. Clearly, agencies from federal to local were unprepared to deal with the multicultural diversity of South Florida (Phillips, Garza and Neal 1994).

After answering the FEMA worker's endless questions and then waiting for weeks, Pat was glad to have received anything to cover her household losses. At one point she was asked to produce three papers documenting her address at the time of the storm. While Pat was fortunate enough to have them, she told of a friend who had lost everything and was having a terrible time qualifying for help. Women applying for FEMA trailers were sometimes needlessly asked if they were married to the partners they were living with, an insult not lost on many. One migrant worker from Guatemala translated the question as, "basically what [they're] saying is, you know, 'Are you a slut living with this guy?'"

Case workers also expressed frustration with the process, complaining of being provided with inequitable, and seemingly inconsistent, screening criteria. After trying for weeks to get repair help for a family without insurance, one case-worker protested, "These people have bent over backwards, I mean they've signed until their arms were cramped . . . we've got three-inch folders; we've got Releases of Confidential Information, and we still aren't one step further in getting their home repaired." Another added, "You see the need, you know the resources are there, but you can't seem to cut through to get them."

Hard-pressed and over-worked FEMA, SBA, and Red Cross home inspectors often missed appointments, arrived late, made only cursory inspections, and left quickly. One of Pat's neighbors described the inspector who finally came to assess her damage and resources as being White, frightened, and in a hurry. "The guy who came to my house – he didn't stay there ten minutes . . . That man, he was just shakin', and I says, 'Sir, are you alright?'" According to our informants, the inspectors often made mistakes when recording the facts about a claimant's losses, further impeding the process.

Direct service providers repeatedly mentioned to us that low-income single mothers were among those having the hardest time. One reason was the extent to which public housing was destroyed and the slow pace at which it was repaired or rebuilt. About two years after the storm approximately 20 per cent of the county's public housing remained unrepaired (Metro Dade Housing and Urban Development 1994). The delay, officials explained, was due to the slow release of public funds, as well as a reliance on private funding for much of the rebuilding. There was also speculation that some officials were hoping that, by delaying or thwarting the release of funds, the projects might be relocated outside their communities. In the meantime, families on public assistance spent months, even years, living in tents, trailers, and damaged apartments. When we conducted our focus group with Pat and her neighbors a year and a half after Andrew, they were still living in apartments with peeling plaster, mildew, broken cupboards, and kitchen appliances rusted beyond use. The repairs at Garden Grove were progressing very slowly, with each family being temporarily relocated to another unit while their own was repaired. Fewer than half of the apartments had been restored and the former daycare center was still being used to store construction supplies.

The women voiced concern about the effects of these disruptions and delays on their children. They described listless toddlers who didn't want to play and schoolchildren who couldn't concentrate on their homework. (For information about effects on the emotional health of children after Andrew, see Tasker 1993; Loudner 1992; LaGreca *et al.* 1996; Jones *et al.* 1993.) Formerly healthy children now had respiratory infections, stomach problems, allergies, and asthma and pervasive nervousness, especially on windy nights. Pat worried that the frustrated mother next door might be beating her toddler who had begun to wet his bed, a concern she relayed to the Health and Rehabilitative Services mental health team canvassing the area. Teenagers in Garden Grove were at a loose end for months, since virtually all recreational and entertainment facilities were destroyed. Even teenagers with a little money had no place to spend it – no music stores, video arcades, pizza shops, movie theaters. The women praised Lion's Club men who recently fixed the basketball court and the two Florida City police officers who were running an afternoon baseball team.

While focusing on their children's needs, it was clear that several of the women were ill. When pressed, they complained of headaches, vision problems, and sore throats attributed to living in damp apartments and a neighborhood still piled high with debris. Because of their heavy responsibilities, women are likely to be exceptionally overworked and emotionally stressed, making it essential to the entire family's well-being that they receive appropriate, accessible, and continuing mental and physical health care. However, Hurricane Andrew damaged approximately sixty health facilities and hospitals and many remained closed for months, thereby forcing people in South Dade to travel long distances under very difficult traffic circumstances to obtain medical care. One advocate for migrant workers said, "I know women whose babies have been sick with high fever and infection . . . [but] they have no transportation. Let's say there is a car. The husband took it to the fields. He couldn't stop working right after the storm."

A number of community-based health care services were eventually initiated. Planned Parenthood set up temporary field clinics at several sites, including the tent cities, when it became obvious that women were having difficulty getting contraceptives (*Miami Herald* 1992l, 8 November). Several agencies brought services directly into isolated communities using mobile vans and equipment. For example, the University of Miami established a clinic at a church in South Miami Heights. The most comprehensive response, however, was the Health and Rehabilitative Services Community Health Teams which, beginning in November 1992, completed a door-to-door canvas of the entire South Dade area, identifying unmet needs and referring victims to appropriate services (Rogers 1993). While most communities eventually received health services, some for the very first time, the need for these services should be anticipated and plans in place to reach victims sooner.

Through first-person and secondary accounts, we learned that women were often the most proactive, outspoken, insistent, and determined of disaster survivors, especially when their families' needs were involved. Her neighbors proudly told the story of how Pat ordered some newcomers to leave when they started drug dealing in Garden Grove. She responded to their praise, "I'd rather die with my shoes on [than let them take over]." We also heard of ways in which Mexican farmworkers, Haitian immigrants, and African-American church women galvanized their neighborhoods into action. Like unpaid community work and family labor, women's social action in the post-disaster period remains a largely unexplored, but very significant, dimension in the analysis of disaster recovery.

Back in Garden Grove, Pat pulls up a chair beside her friends sitting outside to catch the evening breeze and watch the sunset. They swap jokes and make each other laugh as they watch children playing with cardboard boxes amidst twisted trees and crumbling concrete. Though Garden Grove's families still have enormous needs, they have struggled to meet them together and they feel rich in spirit. In Pat's words, "Whatever's been said in the past, we will help each other when there's a crisis . . . and that really made me feel good . . . feel like this was a family. Everybody put everything else aside and just tried to comfort one another. So I think that was one thing we had more than any other community."

Michelle Durant: trailer camp homemaker

We first met Michelle at a storefront Haitian relief agency. She graciously agreed to be interviewed later at home – a FEMA trailer that her family had been living in for nearly two years. When we arrive, she apologizes for the clutter. Every possible space is doing double duty to accommodate two adults (herself and her step-daughter) and four children. As she described it, "The lack of privacy really gets on my nerves. We can't get undressed. It's hard to close the door and there's no room to hang up clothes. My son's bed is in the living area so he can't go to bed until we do."

In a common scenario, the family spent the first few months living in their partially destroyed apartment in a Haitian neighborhood of Florida City, paying rent in exchange for the landlord's unkept promises of repairs. We heard several similar stories, including the plight of a young mother as related to us by her social worker:

> Well, even if you got a [FEMA temporary housing] check, where are you going to go? Now a lot of them, what they did – they make deals with the landlord. OK, we stay, we pay you rent, if you fix. So the landlords are getting the money, but they're not fixing. [She had] no electricity, no lights, and she had her 14-day-old baby, and she was paying $260 rent every month.

When her apartment was eventually condemned, Michelle felt fortunate to have been one of the 3,500 families issued a FEMA travel trailer or mobile home (FEMA 1994). But now, twenty-two months later, the isolated trailer park full of hundreds of crowded, hot, and frustrated residents has become an increasingly ugly place to live. Parks of FEMA trailers throughout South Dade were plagued by crime and violence. According to a FEMA spokesman, the combination of poverty, disaster stress, new neighbors, cramped quarters, and densely packed parks can be explosive (Hartman 1993b). Some camp violence is cross-cultural, or seen as such. In one of our focus groups, Mexican mothers complained that the Guatemalan youths stayed home from school and caused trouble in the camp during the daytime. We heard stories of Haitian and African-American teenagers fighting at the basketball court. The police are called when things get bad, but Michelle says they don't seem to be as responsive as they were at first. Michelle and her children are often frightened by the, now commonplace, sounds of fighting, including domestic arguments, loud bickering between residents, and even gunfire. The fighting in the trailer next door is getting worse each day.

Many of our informants spoke of family conflict as a by-product of the frustration and uncertainty. Overcrowding is no doubt a contributing factor. For example, the forced proximity of children and grandparents with radically different values and standards is likely to cause tension. One Haitian community center responded by increasing their child care program and adding parenting

Plate 7.3 One year after Hurricane Andrew, hundreds of families remain in one of
the trailer parks established as temporary housing
Source: Tim Chapman/*Miami Herald*

workshops to help families learn to negotiate better intergenerational conflicts.
A parent resource center offering respite care and other parenting services noted
an increase in calls for help from middle class parents after the storm. In this
respect Hurricane Andrew was a leveler, sometimes legitimating the act of asking
for help.

As crowded and hot as Michelle's 8 foot by 36 foot trailer is, she keeps her
children indoors as much as possible. She fears the camp is unsafe. "Since we
are into the camp, every week there are shooting or stabbing, killing, robbing.
They rob my trailer. They stole my [license] plate too." She and the children
find it hard to sleep because of the crescendo of voices and music coming from
the group of men who hang out each night by the camp's only working
telephone.

In this culturally mixed camp, the common denominator is poverty. The
primary focus is on survival and the displaced families tend to keep to themselves. As she peers out from behind the curtains framing the trailer's window,
Michelle spots her neighbor sweeping out her trailer. She wishes again she knew
enough Spanish to talk to the woman. But only the young seem to mingle in
the camp, and then often not peacefully.

We have been talking quietly, since Michelle's step-daughter Denise is asleep, tired after a long shift as a waitress. The cafe where she works is doing a booming business serving out-of-state construction workers. Daniel, Michelle's teenage son, spends his days searching for construction work. Though the newspapers are filled with pictures of White men on temporary construction jobs, he has been unsuccessful in finding a job. Michelle lost her housekeeping job when Hurricane Andrew destroyed her employer's home and she has not been able to find another. There is little demand for domestic help since so many households left the area. If she had transportation, Michelle says, she might be able to find work to the north, but the family's vehicle was crushed by a falling tree. After the hurricane, Michelle's husband Yves began to drink heavily and became abusive. He eventually deserted the family, leaving behind three children from his marriage to Michelle and two from a prior marriage. Food stamps and Denise's income have supported the family since he abandoned them and Michelle lost her job.

Yves' response to post-hurricane life was not unique. Though the reasons varied, male desertion was frequently reported among our respondents. Anne, an Anglo woman we interviewed while she was living with her two small children at a battered women's shelter, had this to say about her partner: "He couldn't take the pressure, being used to everything, and then coming down to no eating because we could not find food . . . And then he was beating me up . . . He really went crazy." Anne hopes that her application for FEMA benefits will be processed before her husband's so the check will be issued in her name. She plans to spend it on bus tickets out of Miami, but she's afraid he may get the money instead.

Our interviews with service providers confirmed that the first person from each address to submit an application – most often the man with transportation – was usually the one who received the check. There were many reports of FEMA benefits intended to replace household possessions or to provide temporary housing being misused by men for personal purposes, such as buying cars or supporting relatives in other countries. Even if she got the check, there was no guarantee she could keep it. According to a social worker, "There is conflict, more conflict. The woman gets the money. It's to replace her furniture, but he says, 'No, it's our money. Give it to me.' And he takes it any way he can."

While urban planners and politicians envisioned gracious planned communities rising Phoenix-like from the urban wreckage, rebuilding in South Dade proceeded haltingly. Recovery efforts concentrated on owner-occupied single-family homes – dwellings beyond the reach of many of the area's long-term residents. Although FEMA encouraged those remaining in the government trailers to buy them at low cost, even the pad rental and hookup fees were unaffordable for many. Our work affirms that in the painstaking process of recovery, those with the least resources before the storm – often single mothers and grandmothers – tend to get stuck in the limbo of "temporary" housing. When disaster strikes an area heavily populated by low-income families, their margin of survival and independence is already very slim. Gender, race, culture, and social class clearly shape women's short-term needs, as well as their long-term prospects for recovery.

Michelle and her family are getting by, although life is hard. She feels isolated at the remote camp, cut off from the services her family needs. With over 300 trailers there, she asks if we know why some of the health and social agencies couldn't have been housed in the park instead of miles away. But it is the uncertainty of the future that worries her most. She wonders how her family will get to the grocery store and the health clinic when the FEMA-funded jitney stops; how long she can keep her teenage son out of trouble in the camp; where she will get the money for school supplies and clothes for the younger children; when the man from FEMA will return to collect rent for the trailer; and if the family will ever find an affordable apartment large enough for the six of them.

As we prepare to leave, Michelle asks for a ride up to Florida City. When she has a chance, she goes there to purchase groceries and to visit friends at the community center where they sort donated food and supplies. Sometimes she pitches in to help and, in exchange, takes home enough to make a good meal for the children. As the sun heats up the tiny trailer, Michelle quickly gathers her family for the unexpected ride out of the camp on yet another sweltering day in Miami.

Elena Moreno: businesswoman

We met Elena in her home office, where crowded stacks of paperwork left an impression of controlled chaos. She is a striking, middle-aged Cuban-American with three adult children, all living in the Miami area. Robert, her husband of 27 years, sells real estate. After many years helping in his business and rearing their children, Elena now owns and manages her own small business. Their suburban home also shelters Elena's elderly parents, who evacuated there the day of the hurricane. Elena took on much of the hurricane-related work to help both her parents and her adult children's households. Because her family considered Elena's time more flexible, and perhaps less valuable, it was she who waited in the long lines for hurricane supplies; she who made contingency plans with each of her adult children's households; she who kept distant kin informed before and after the hurricane. Since the storm, Elena has been at the center of the recovery process of the five households of her extended family.

Because "kin work" falls largely to women across all social classes and cultures, initiating family support networks in time of disaster is usually their responsibility. Some of Andrew's victims sent their children out of Miami to be cared for by grandmothers and aunts until the crisis was over; others took in relatives and foster children. There were numerous accounts of the giving and receiving of material and emotional help among "family" – whether fictive kin, child care providers, collective households, or relatives from other households. The degree to which women in disaster conditions are available to take on these tasks, in addition to their varying combinations of employment and informal sector work, clearly impacts on household recovery. Qualitatively different from replacing roof tiles or clearing debris, women's "invisible" domestic work continues unabated or increases in the wake of disaster. In tent cities, in makeshift camps, in trailers, and in homes half destroyed or under construction, we witnessed women cooking, cleaning, washing clothes, changing diapers, and comforting distraught children and elderly parents.

Elena's parents' home was virtually destroyed. When they applied for assistance they found themselves in a *Catch 22* situation. Their FEMA application was denied because they had insurance. Yet, they have not collected. In fact, they haven't even seen an insurance adjustor. It is rumored that their small, local insurance company is one of many on the verge of bankruptcy. Lacking the savings to cover the cost of rebuilding, the elderly couple continue to live with Elena and Robert. Elena's daily obsession was trying to determine the status of her parent's claim. With raw anger in her voice, she describes her innumerable and unsatisfactory phone calls to the insurance company, FEMA, and other agencies on her parents' behalf.

Women's domestic roles typically extend to tasks linking families with agencies, schools, churches, and other organizations. Bureaucratic negotiation and family advocacy became part of their daily routines after Andrew. The difficulties of expediting relief claims through the complex application and qualification procedures were further complicated for non-English speakers, recent or undocumented migrants, mothers of small children, and the elderly. The demands put on relief organizations were unprecedented and it was inevitable that some of those turned away would be disgruntled. However, among the women we spoke with there was a persistent feeling that many decisions were unjust, whether intentional or merely the result of disorganization or short-sighted policies. Not having to fight the battles alone, but drawing on the advocacy and emotional support of other family members, seemed an important factor affecting recovery (see Chapter 8). Dealing with relief organizations was not uniquely a female task, but it seemed that it was primarily women who returned again and again to fight for help for their families. As reported by one long-term caseworker, "It's been the woman that pulled the family through, and it's the woman that continues to hold it together." Social class, racial/ethnic, and cultural patterns no doubt cross-cut this generalization and warrant additional attention as researchers explore gender relations in disaster-stricken households.

Elena's parents are sleeping in the living room so that Elena can use the spare bedroom as her home office. The house was even more crowded during the first few months when Elena kept her three young grandchildren each day while their parents worked to make their own damaged houses inhabitable. The neighbors helped Elena with child care, just as they helped in so many other ways since the storm.

Starkly apparent when normal routines are disrupted by crisis, many ordinarily invisible caregiving networks are critical to family and community recovery. In our focus group with family daycare providers, we learned how many of them continued to keep children even though their own homes and lives were in disarray. Often these services made the difference between parents keeping or losing their jobs. One provider told of parents dropping children at her door, along with a bag of ice, just days after the hurricane. She had no electricity, running water, or telephone, but the parents needed to get to jobs they could not afford to jeopardize them. Many of these caregivers were very attached to their charges, functioning much like mothers and grandmothers. One elderly African-American told of making her way through blocked city streets and fallen electrical lines to find the three young children of a single father she feared would not cope well after the hurricane. As she had anticipated, the father was unable to provide adequate care. She took the children in to live with her for six months until their mother came to Miami to get them. However, many who formerly provided care to children, the ill, disabled, or elderly were unable to continue their services due to Andrew's impact on their own lives, thereby compounding the problems of those relying on them. Disaster planners must be aware of the effects of the loss of these informal caregiving networks and consider how their services can be supported or replaced during the emergency and recovery periods.

Elena was fortunate to be able to care for her parents and still keep her business operating. She proudly explained that her husband took over virtually all of the tasks of restoring their home while she concentrated on her business. It was important to rebound quickly and keep her six workers employed. After assessing the damage, they decided to disperse various business operations and supplies, using staff members' spare closets and garages. On many days, they all worked directly from Elena's tiny bedroom office, testing their patience and adding further commotion to the household. In her usual fashion, Elena focuses on the positive, "six people working in a space about this big [she points to a small bedroom] for all the time the reconstruction was going on. It was real hard, but then when we were ready to move into our own offices, everyone was all crying because they'll miss each other."

For Elena the support she received from the local chapter of a national organization of professional and small business women was critical. Its monthly meetings, which resumed only a month after Andrew, were opportunities to talk about hurricane experiences and feelings as much as about business recovery. Elena says, "When you have a situation you haven't seen before you know its not the first time it's happened. Someone else has had it happen to them. So who do you talk to? Other business owners. But, in the context of a business group it turns out to be a support group." Through this network, women were helped by other chapters. When a telephone poll revealed that the South Dade members needed computers and office equipment, but also shoes for their children, they got both. Out of this experience the national association started a revolving cash fund to be used to assist sister chapters in future disasters. When considering women's roles after a disaster, their occupational or professional resources are likely to be overlooked, yet these formal and informal networks are an important resource in long-term community recovery.

Like her husband, Elena is also active in her church and in several community organizations. She smiles broadly when asked about her work with Women Will Rebuild. When it seemed that economic redevelopment was taking precedence over the more immediate needs of South Dade families, a coalition of women's groups formed to protest. The emergent coalition, Women Will Rebuild, eventually included women from over forty religious groups, youth organizations, feminist networks, social and professional groups, and service organizations, representing all of the major racial and ethnic groups of Miami's diverse population. These women challenged We Will Rebuild – the powerful group of business and community leaders responsible for distributing millions of donated and public dollars. Women Will Rebuild lobbied for more funds targeting the needs of local women and children. They also developed and publicized a multicultural roster of experienced women leaders appropriate for appointment to We Will Rebuild.

As a representative of her businesswomen's association, Elena began attending meetings three months prior to our interview. Some of her Hispanic friends remain skeptical of the Anglo-dominated group's use of feminist language and a consensus decision making process. And Elena questions how many more precious Saturday mornings she can give up in such a hard year. But she wonders who except Women Will Rebuild speaks for low-income families and for funds for programs to serve women and girls. She also wonders when Miami has ever seen such a multicultural coalition, male or female. Although the coalition evidently dissolved simultaneously with We Will Rebuild, many believe its very existence left a legacy of effective lobbying and coalition work in a divided city, showing that a wide range of ethnically and socially diverse women can galvanize to action over shared concerns.

As a daughter, mother, and partner, Elena has carried a heavy family load throughout this crisis while struggling to keep her business operating. To accomplish all this, she has drawn heavily upon the strength of her family, neighbors, and women friends. When Elena told us of the gains that have come with the struggle, including a new-found assertiveness, new interests, and new friends, she spoke for many women for whom the hardships of Hurricane Andrew afforded opportunities for personal and professional growth.

CONCLUSIONS

These composite profiles document the range and commonalities of women's experiences, illustrating that women are pivotal in the intersection between household and community recovery. While their needs and experiences are in many respects gender specific, as well as deeply influenced by class and ethnicity, they also provide critical insights into neglected, yet central, problems, processes, and mechanisms of household and community recovery. We conclude that a gendered analysis is crucial to understanding and mitigating against future impacts of disasters on families and communities. Research and theory that more fully account for the experiences and needs of women will generate more complete and accurate knowledge, yielding more effective policies for future disaster management. Toward this end, we conclude with a series of research questions and policy recommendations.

WOMEN, GENDER, AND DISASTER: A RESEARCH AGENDA

We draw on this study of women and Hurricane Andrew to identify key directions for future research on gender relations and disaster. Recognizing how intertwined the issues are in actual experience, here we distinguish analytically between questions raised by first, women's vulnerability to disaster, second, the gendered impacts of disaster, and third, women's capacities and resources for responding to disaster. These questions can and should be addressed at different levels of analysis. Issues raised by gendered disaster impacts, for example, might be investigated at the individual (the vulnerability of elderly women living alone), household (change in the gendered division of labor), organizational (new service demands and responses for organizations serving women), or institutional levels (post-disaster changes in the female labor market). They can and should be addressed in relation to cultural and economic patterns, for example culturally specific disaster impacts on recent immigrant women and economic patterns in women's recovery capacities. Each is also framed by global patterns relevant to hazard and disaster, such as the increased prevalence of female poverty and household headship.

Disaster vulnerability

- How do physical, economic, and social conditions of life place women at special risk from disasters, for example pregnancy and reproductive needs, longevity, caregiving responsibilities, poverty, head-of-household status, nursing home or public housing residence? What specific conditions produce varying levels of vulnerability for women in differing life circumstances?
- What regional or local patterns increase women's vulnerability to disaster, for example farm labor migration patterns, immigration trends, women's formal and informal economic activity, availability of affordable housing, rates of violence against women?

Disaster impact

- What models can be identified in disaster-struck communities for responding to women's needs, for example for personal safety, economic security, and mental health services for caregivers?

136

- How are at-risk groups of women most effectively reached with preparation, evacuation, and mitigation information, as well as recovery services? What media effectively deliver information to target groups?
- What patterns of bias (gendered, economic, cultural) can be identified in the practices of responding agencies in the private and public sectors? How are these maintained informally? Under what conditions are they most effectively challenged?
- How are women's particular needs identified by disaster planning and response agencies and emergent groups? What organizational conditions increase the visibility of gender issues?
- What training programs or other mechanisms best sensitize paid and volunteer disaster workers to gender bias? How do disaster management agencies address gender issues in program planning, implementation, and evaluation?
- How distinct are the long-term recovery problems (emotional, material, financial) and coping mechanisms adopted by women and men? What factors sustain these patterns and how?

Disaster response capacity

- How have women in various settings (rural/urban, developing/developed societies) organized to meet household, neighborhood, and community immediate post-disaster needs? What conditions or resources facilitate women's effective mobilization around gender issues in disaster?
- Under what conditions are women's formal and informal networks and community leadership effectively identified and utilized at all stages of community disaster response?
- What cultural patterns make women effective disaster responders at the household and community levels? What patterns erode women's capacity for self-recovery?

WOMEN, GENDER, AND DISASTER: POLICY CONSIDERATIONS

What lessons are there for policymakers in women's experiences related to Hurricane Andrew, as we understood and portrayed them in this chapter? As our composite profiles suggest, women absorb the social costs of being largely excluded from disaster planning, and response and recovery initiatives. We conclude by urging a redirection of disaster planning to account better for the impact of gender relations on social structure and interaction. This is the key, we believe, to disaster planning and response which more effectively meets the needs of women. Again we pose the issues broadly, recognizing the wide range of groups, agencies, and organizations engaged in disaster initiatives.

Organizational practice and culture

Informal practices in disaster organizations contribute to a "gender-blind" approach which, in practice, disadvantages women. Historically male-dominated and grounded in military culture, disaster organizations should evaluate their routine operations for potential sources of gender bias.

Toward this end, we recommend that disaster agencies:

- undertake in-house evaluation of employment practices, i.e. recruitment, hiring, training, assignment, promotion, retention;
- afford opportunities and support for female/male staff and volunteers in non-traditional positions;
- establish gender-sensitive program planning, implementation, and evaluation;
- institutionalize comprehensive and on-going gender-sensitive training to help managers, staff, volunteers, contractors, and others recognize covert and overt patterns of bias.

Preparation and mitigation

Because women are instrumental in preparing households and kin for disaster, they are essential actors in community-based disaster planning and local mitigation initiatives. They must be fully engaged as equal and active partners in order to build democratic disaster-resilient communities.

Disaster planning groups in the public and private sectors should:

- analyze the vulnerabilities of women locally, and proactively include targeted groups of women in community education and other programs designed to mitigate hazards, reduce disaster impacts, and promote community recovery;
- include in planning initiatives key representatives from relevant women's community groups, for example women's health services, women in construction, business and related professional associations, public housing tenant associations, and advocates for homeless, migrant, and battered-women's services;
- plan ways to assist particularly vulnerable groups of women in preparing and/or evacuating their homes, for example public housing residents, single mothers and women living alone, senior and disabled women, low-income women, women whose migration status or language skills may exclude them from traditional networks;
- actively build on existing women's groups and leaders, whether formal or informal, to access women otherwise not likely to be targeted by traditional warning and information media or by neighborhood preparedness campaigns;
- undertake community profiling and assessment which includes gender-specific vulnerability indicators, for example proportion of single-headed households, average female wage and local rent costs, predisaster demand for domestic violence services, migration and immigration patterns.

Emergency response

Women are symbolically central to disaster relief as the helpless victim tearfully grasping the strong male hand of the rescuer. Yet women's particular needs in the immediate aftermath of disaster are rarely addressed.

Formal disaster agencies, as well as established and emergent groups serving survivors, will better serve women if they:

- broaden their target populations to include those groups of women identified as most vulnerable to particular hazards;
- organize relief centers to ensure that they meet women's needs for child care, translation services, access to public transportation, and have staff sensitive to cultural and economic differences in household structure and power relations;
- examine policies and practices in relief agencies for possible gender bias, for example head-of-household regulations disadvantaging women heading collective or multi-generation households;
- organize emergency housing to ensure that it meets women's needs for personal safety, public transportation and telephone services, accessible child care or respite care, reproductive health care, and gender-sensitive mental health services;
- actively search for ways to ease the burdens of disaster responders, particularly those from the local community, who are typically women with heavy domestic responsibilities, such as providing mental health counseling, flexible work schedules, and child care.

Long-term recovery

Perhaps the greatest challenge to disaster planners is responding to the long-term needs of particularly hard-hit women. Economic redevelopment initiatives, for example, rarely consider women's economic status or needs outside of the family economy. As illustrated throughout this book, predisaster inequalities and vulnerabilities are reflected in long-term recovery problems and, thus, low-income and poor women, particularly among minorities, have a particularly difficult time.

Planners knowledgeable about gender relations in family and community life will design recovery strategies useful to both women and men. To this end, we urge them to:

- fully engage the resources of local women leaders and their informal and formal networks;
- anticipate the need to replace local social services volunteers in heavily impacted areas and to augment services in such areas as domestic violence, reproductive health, child care, and elder care;
- anticipate that low-income women heading households will be particularly slow to recover and have longer needs for housing and other forms of recovery assistance;

- look for ways to support the recovery of multigenerational households and to help extended families with the recovery of their most vulnerable members;
- monitor reconstruction to ensure that women's particular needs are identified and addressed, for example in economic redevelopment plans or the distribution of donated labor and materials;
- plan for the housing and safety needs of relief and reconstruction workers coming into the stricken community.

Throughout the unfolding process of disaster, both the short-term needs and the long-term interests of women must be accommodated if response is to be truly effective – but, first, they must be better understood. Disasters do not impact women and men in uniform ways, nor all women uniformly. Analyzing the specific needs of women in varying life circumstances is a planning task for disaster managers and a practical challenge for responders. Adopting a gendered perspective is the first step toward implementing the kind of disaster response Miami women so clearly needed.

8

STRETCHING THE BONDS
The families of Andrew

Betty Hearn Morrow

Do you think that if I had family I'd be here in this tent?
Honduran resident of Harris Field Tent City (translated)

The havoc raised by natural disasters extends far beyond physical destruction. Hurricane Andrew resulted in surprisingly few serious injuries and deaths, especially considering its strength and landfall in a large urban region. The human suffering caused by disrupted lives and relationships, however, has been extensive. The destruction of homes, businesses, and neighborhoods blew away the infrastructure on which the daily lives of hundreds of thousands of people depended. The initial shock and chaos was only the beginning of weeks, months, even years, marked by tremendous demands, changes, and emotional upheaval.

Disasters of this magnitude have particularly profound effects on victims' most intimate social group – their families. The demands of rebuilding homes and lives place tremendous strains on a family. Many are unable to cope and are torn apart, while others persevere and even grow stronger. Some are able to obtain important support and assistance from relatives. In other cases, these kin networks have little to offer or do not exist. This chapter focuses on the importance of kin networks in helping families deal with the multitude of tasks and activities resulting from Hurricane Andrew, but also examines some of the stresses created within families by this experience.

The diversity of Dade County's population affords a unique opportunity to study families varying in socioeconomic status, culture, and household living arrangements. This chapter brings data collected from a number of FIU Disaster Research Team projects to bear on family-related issues. In addition to the tent city interviews, the FIU Hurricane Andrew Survey, and the Family Impact Project, two other studies yielded relevant family data. They are the South Miami Heights Study of households in a severely impacted, multiethnic, working class neighborhood conducted one year after Andrew, and the American Red Cross Project in which we interviewed social services providers. Reports and documents from public and private agencies and research groups, as well as articles from the print media, also added to our understanding of family response to this disaster.

FAMILIES IN DISASTER RESEARCH

Studies of the dynamics of disaster preparation, evacuation, and recovery often use the household as the unit of analysis (cf. Bates and Peacock 1993). This is justified on the grounds that it serves as the primary functional unit of society. Regardless of the larger economic or political system, people throughout the world typically react to, experience, and cope with crisis events as members of households. Most households are family units; that is, the people living there are bound together by a set of primary relationships and obligations. In some sense, the shift from household to family is a subtle one. And yet it can have very real qualitative consequences because now both *instrumental* and *expressive* functions become critical for understanding recovery. The importance of family contact, stability, and networks to the individual's sense of normalcy following disaster cannot be overemphasized (Quarantelli 1960). Disasters need to be understood as family crises in which these emotional bonds are critical factors for recovery (Hill and Hansen 1962). Recent psychological research, for example, has linked post-disaster stress and trauma to the breakdown of support systems (Kaniasty and Norris 1993). Households are often embedded in non-residential familial networks that can also be called upon in times of crisis. Thus, the shift from household to family can result in qualitatively different questions and issues.

Our approach utilizes an ecological perspective that examines the nexus between a family and its larger environment (Hewitt 1983; Faupel 1985; Bates 1993). Contingency and negotiation characterize the linkages between a family and the countless semi-autonomous social units operating within its competitive social field. The ability of a family to respond to disaster – its openness – will be contingent upon its internal dynamics, as well as its perceptions and definitions of the situation (Drabek and Key 1976). If a family does not "see" the impending or potential disaster, it will not act. Equally as important will be a family's ability to marshal internal and external resources. If access is closed or blocked, then the family is at risk. Factors that alter the nature of the contingency relationships by either increasing or decreasing the probability of gaining access to resources will be critical. The physical condition of the community or broader field is another important factor. When the infrastructure has been extensively damaged, the everyday activities of family life become extremely difficult and can seriously hamper individual and family recovery (Trainer and Bolin 1976).

One area which has received a good deal of attention is the response of families to advance-warning disasters (Nigg and Perry 1988). It is within the family that most people "define the situation" and make decisions about household preparedness and evacuation (Drabek and Boggs 1968; Fitzpatrick and Mileti 1991). Family characteristics such as ethnicity, class, age of members, and internal power relationships influence warning compliance and action (Taylor 1978; Perry and Mushkatel 1986; Perry 1987; Perry and Lindell 1991). Relatives' homes tend to be preferred places of refuge, particularly among poor and minority families or those with children or elderly members (Quarantelli 1960; Drabek and Boggs 1968; Bolin 1982; Perry and Mushkatel 1986).

In the highly competitive post-disaster recovery period each family's social and economic position and connections within the larger community can be critical factors influencing the outcome for their household (Drabek et al. 1975; Bolin 1982; Trainer and Bolin 1976; Drabek and Key 1982). The family unit has been likened to an octopus extending its tentacles outward to connect with other social units (Drabek and Key 1976). Some have stronger tentacles prior to a disaster and are, therefore, able to secure more help afterward. Whether called tentacles, "social webs" (Drabek et al. 1975), "defense in depth" (Hill and Hansen 1962) or "institutional and kinship embeddedness" (Bolin 1982), these connections can be used to promote recovery.

Comparative disaster studies have revealed three principal modes of family recovery: first, the autonomous use of personal resources (such as savings or insurance settlements); second, reliance on informal kinship support systems; and third, the utilization of institutional resources (such as government assistance). While victims often make use of all three, the extent to which one dominates is primarily determined by the larger political and economic setting (Dynes 1975; Bolin and Trainer 1978; Bates and Peacock 1989b). In other words, contextual factors will strongly impact the mode, rate, and level of recovery. In modern industrialized settings, kinship assistance is less apt to be the primary source of help, but it is still important (Fogelman and Parenton 1956; Erickson et al. 1976; Drabek and Key 1976; Bolin 1982; Nigg and Perry 1988). The economic and personal resources of a family, including its relative position within the community power structure, will determine the extent to which it can autonomously facilitate its own recovery.

The interpersonal dynamics within family units provide important measures of the social and emotional effects of a disaster. Family roles and responsibilities are likely to undergo considerable change associated with household and employment disruption, economic hardship, and the demands of recovery. As life patterns are disrupted and family relationships change, role frustration is likely to occur (Bates et al. 1963). When agencies have to be relied upon to provide basic needs, family provider roles can be undermined.

Experiencing a major disaster is, of course, a stressful event. For many, the frustrations of trying to cut through the bureaucracies of insurance companies, mortgage companies, social services agencies, and various governmental entities were enormous and long-lasting. Because of the extent of the destruction and remote location of South Florida, even for those with resources, the acquisition of the necessary materials and labor to repair and rebuild their homes was extremely difficult. It is not surprising that stress levels were high, even among children. The findings of various studies of the psychological effects of the storm on children included: over half of one sample exhibiting symptoms of moderate to very severe levels of post-traumatic stress disorder (PTSD) three months after Hurricane Andrew (Vernberg et al. 1996); a correlation between PTSD and the extent to which children's homes were exposed to the hurricane's severity (Parker *et al.* 1997); and elevated levels of stress among elementary children six months post-disaster (Jones *et al.* 1993). When the sample from the Vernberg *et al.* (In press) study was evaluated 18 months later, one-third of the children were still manifesting PTSD symptoms (LaGreca *et al.* 1996). While there were no significant differences in reported levels of psychological stress among college students who experienced Hurricane Andrew, they showed more symptoms of physical stress and did more poorly on cognitive tasks than classmates who had not experienced the full force of the hurricane (Rotton *et al.* 1996).

The effects of disasters on marital relations are unclear. Erikson (1976) reported testimony from survivors of the Buffalo Creek flood about changes in their marital relationships – most often decreased levels of emotional and sexual intimacy. It is common for divorce rates to fluctuate after a disaster, eventually returning to previous rates (Friesema *et al.* 1979; Morrow 1992). Lasting effects tend to reflect predisaster relationships; in other words, strong bonds get stronger while tenuous ones become weaker or ultimately break under disaster stress (Drabek and Key 1976). There is some indication that going through a disaster recovery experience may negatively impact relations with non-household kin (Trainer and Bolin 1976; Erikson 1976).

While these discussions give some idea of how families function in disaster, there is much we do not know. For example, many aspects of the associations between family attributes – such as ethnicity, class, and family structure – and disaster-related decisions and outcomes remain largely unexplored. While the greater difficulties faced by minority and poor families at all levels of response have been well documented (Hill and Hansen 1962; Cochrane 1975; Perry and Mushkatel 1986; Bolin and Bolton 1986; Perry 1987; Phillips 1993b; Blaikie *et al.* 1994), much remains to be learned about the ways in which these factors are reflected in family responses and recovery outcomes. Hurricane Andrew provided a rare opportunity to study a large number of families – diverse in structure, class, and ethnicity – as they struggled to recover from a major community-wide disaster.

THE FAMILIES OF ANDREW

Hurricane Andrew impacted about 130,000 households in South Dade – about three-fourths of which were family units of two or more related persons (Metro Dade Planning Department 1992). More than half of the victim population was composed of what traditionally are termed minorities – mostly Hispanics and/or Blacks. The average household size was approximately three persons and the households were structurally diverse. Non-nuclear kin often lived together. For example, at the time of the 1990 Census, over 10 per cent of Dade County homes included grandchildren of the head of household. Almost 15 per cent of the family households were headed by a woman, with much higher rates among the poor. There is also a great deal of variation in social class. While there are expensive homes scattered throughout the area, in general, as you move southward, the upper-middle class communities near Kendall are gradually replaced by working class neighborhoods, with an increase in multiple-family rental units.

Extended family networks are the norm in South Florida. According to the Hurricane Andrew Survey, about 75 per cent of all Dade County households have at least one relative living in the area, and 45 per cent have more than five. Black and Hispanic residents are significantly more likely to have kin living nearby, as are nuclear families (parents and children) and female-headed families. Thus, the families in Andrew's path provided a unique opportunity to study the role of kin networks in disaster preparation and response.

FAMILIES AS NETWORKS OF SUPPORT

Before the storm

As discussed in Chapter 4, most of the households in the Hurricane Andrew survey received their hurricane warning and preparation information from television. However, about one-third said relatives were a very important information source. As found elsewhere (cf. Perry and Mushkatel 1986), this was particularly true for minority households. Families with children, especially female-headed, were also more apt to have talked with relatives about preparing for the storm.

While information is important, we were also interested in the extent to which the giving and receiving of help occurred within kin networks. Somewhat surprisingly, only 14 per cent of the households in the survey reported receiving assistance from relatives when preparing their homes and, among those reporting having kin in the area, the rate was still only 16 per cent. Several logistic regression models predicting the log odds of giving or receiving help were developed to determine if there were significant socioeconomic and ethnic differences in the utilization of kin networks.

We begin by focusing on the receipt of aid from relatives. For this analysis, Black and Hispanic households are compared to Anglos. We might expect several factors related to need to influence the receiving and giving of aid. For example, households less able to help themselves, such as elders, widows, and single mothers, should be more likely to have received kin help. Homeowners and those living in evacuation zones would have more preparation work to do and, hence, would be more likely to need assistance than would renters and households not expected to evacuate. The presence of kin in the area would, of course, be a critical factor in all of the models. Measures for each of these factors are included in the model predicting whether relatives helped presented in Table 8.1.

Table 8.1 Logistic regression models predicting relatives as source of help preparing for Hurricane Andrew

	Relatives helped	*Helped relatives*
Constant	−3.006	−3.391
Have relatives in area	0.570[c, d] 1.767	1.398[a] 4.048
Black	0.602[b] 1.827	0.690[a] 1.994
Hispanic	0.582[b] 1.789	0.291 1.338
Income	−0.054 0.947	0.005 1.005
Single adult		−0.040 0.961
Couple		0.203 1.225
Elder household	−0.177 0.838	
Widow	0.890[b] 2.436	
Single mother	0.377 1.457	
Own home	0.727[a] 2.068	0.439[b] 1.552
Evacuation zone	−0.205 0.815	0.400[b] 1.492
Chi Square	39.657[a]	49.727[a]

Source: FIU Hurricane Andrew Survey, n=1,024
Notes: a p≤.01
   b p≤.05
   c p≤.10
   d The first number refers to the unstandardized logistic regression coefficient and the second is its antilog.

The results indicate that minority families are more apt to have been helped by relatives. The odds of Black and Hispanic households receiving kin help, at 1.827 and 1.789 respectively, are nearly twice as high as those of Anglos. Contrary to previous research, income does not have an effect. While elder and single-mother households display no important differences, the odds for widows are 2.4 times higher. As anticipated, homeowners are almost twice as likely to have had assistance as are renters.

A similar picture emerges when examining households providing aid to their relatives. About 18 per cent claimed to have played a major role in preparing a relative's home for the hurricane. This figure changes little, to only 22 per cent, when focusing on households with relatives in South Florida. The second model in Table 8.1 was developed to explain a household's odds of aiding relatives. The model is essentially the same except that single adults and couples without children (i.e. households without dependents) are compared to others, anticipating that they might be more available to help. We also expected that homeowners and those living in evacuation zones would be less likely to have time to assist their kin. The results once again suggest that income is statistically insignificant, though ethnic differences appear. Black families are more likely to have helped non-residential family while Hispanics are similar to Anglos. There are no important differences in the model related to household composition. The expected tendency for single-adult and couple households to be helpers does not materialize. Surprisingly, homeownership and living in an evacuation zone actually increase the odds of helping other relatives. It may be the case that having to think about protecting one's own dwelling promotes more awareness of the needs of others, particularly those households to which they may have evacuated.

Overall, kin networks appear to be under-utilized during hurricane preparation: While nearly 75 per cent of respondents had relatives living nearby, less than 20 per cent reported assisting or being assisted by them. Few people reported receiving any external help, and if they did, it was most likely from relatives, especially among minority families. There were considerable ethnic/racial variations in both giving and receiving help from familial networks. Black and Hispanic households were more likely to have received help and Black households were more likely to have given help. Income, per se, did not play a role, though homeowners were more likely to help and receive help.

During the storm

As discussed in Chapter 4, our data support the claim that families tend to evacuate as units (Perry and Lindell 1991). About 54 per cent of all households located in evacuation zones evacuated entirely, with only an additional 5 per cent leaving at least one household member behind. Only about 4 per cent reported staying because they had nowhere else to go, but this increased to 7 per cent among the poorest respondents. While the evacuation of coastal areas and trailer parks went surprisingly well, we interviewed households left without transportation. A particularly poignant case was a public housing resident who described her efforts to nail plywood over her windows to protect her apartment. When she was ready to leave, she could find no transportation and was forced to walk with her three young children to the nearest highway and hitch a ride to a shelter.

Most South Florida residents waited out the storm in private homes. Often relatives and friends congregated together in what they perceived to be the safest house or apartment. About 27 per cent of our total sample reported having relatives stay in their home during the storm and the rate was significantly higher for Hispanics. One Puerto Rican woman remarked, "I wanted to be with my family. I said, well, if I died, something happened to us, I want to die close to my family." While some people were very frightened before the storm, most later felt they had greatly underestimated it.

Literally thousands of Andrew's victims endured several hours of moving from room to room during the night as their home was blown apart, piece by piece. Whether a result of the progressive destruction of their houses forcing them into safer locations or simply following instructions of radio and television advisories, tens of thousands spent the last hours barricaded in bathrooms and closets. As vividly described by an African-American grandmother:

> We went into the bedroom. It was like the devil demon was behind us ... The walls started shaking in there. I said we better go in the closet. So we put the little children in first. Then the top of the closet went and stuff started falling on us. [There were] 13 of us in the closet and 2 of them were pregnant.

Another said:

> We could hear cars flipping over in the parking lot, all the trees coming down and the roof in my Mom's room cracked open and it started raining inside the apartment, so we all [16 adults and children] went into the bathroom ... It was terrible and we were really dying of heat ... It was hard to breathe.

And another: "It took the roof. When we saw that, everybody got scared: This is it. So that is why we went into the bathroom." Several parents told of kissing their children goodbye as they placed them in the bathtub and covered them with mattresses. When the long night was over, these families had survived an ordeal which many felt left them more appreciative of each other and which would become part of family folklore in the years to come.

One of the first things many did after the hurricane passed was to attempt to locate relatives living in the area. This proved to be an extremely difficult task since telephones were not working and roads were blocked with debris. An elderly woman interviewed in a tent city recalled how she picked her way through the ruins of her house, but could not find a pair of shoes. She walked barefoot for four blocks through the wreckage of her neighborhood to her sister's house to check on her.

After the storm

Assistance from relatives was somewhat higher after the storm. According to our survey, 24 per cent of those with relatives in the area reported receiving major kin assistance with such things as supplies, debris removal, and repairs. Similarly, 30 per cent of those with family in the area reported assisting relatives after the storm. Once again, a set of multivariate logistic models predicting the log odds of receiving and giving help were developed. In addition to the variables used in the pre-hurricane assistance models, we added whether the respondent resided in South Dade and had a home that sustained major damage. In the first model in Table 8.2 predicting the odds of receiving kin help in the aftermath, family and household attributes are not significant. In addition to having kin in South Florida, the only significant predictors of receiving aid are if the home sustained major damage and was located in South Dade, where most of the serious destruction occurred.

Table 8.2 Logistic regression models predicting relatives as source of help after Hurricane Andrew

	Relatives helped	Helped relatives
Constant	−2.673	−3.741
Have relatives in area	0.776[a]	2.049[a]
	2.022[c]	7.759
Black	−0.247	0.484[b]
	0.781	1.622
Hispanic	−0.126	0.053
	0.881	1.054
Income	−0.013	0.190[a]
	0.988	1.209
Single adult		−0.194
		0.824
Couple		−0.294
		0.745
Elder household	−0.121	
	0.886	
Widow	0.290	
	1.336	
Single mother	−0.192	
	0.826	
Own home	0.089	0.127
	1.094	1.136
Live in South Dade	0.685[a]	−0.075
	1.985	0.928
Sustained major damage	0.486[a]	−0.002
	1.625	1.002
Chi Square	97.434[a]	112.814[a]

Source: FIU Hurricane Andrew Survey, n=981
Notes: a $p \leq .01$
   b $p \leq .05$
   c The first number refers to unstandardized logistic regression coefficient and the second is its antilog.

149

A different picture emerges in the model predicting whether the household helped relatives. Black families are about 1.5 times more likely than Anglos to have provided help. Interestingly, for the first time income differences are significant: the chances of having been a helper increases markedly with the respondent's income. These findings suggest that under severe conditions family networks become an important source of help in the aftermath of a disaster. This point is further supported by the observation that, while only 21 per cent of the total Dade County sample received kin help, the rate increases to 44 per cent among the sub-sample from South Dade with relatives in the area. If the Miami area is any indication, urban families tend to have relatives living in the community and these networks can be expected to provide help to about one-quarter of an affected population – more when the need is extreme.

Sharing losses and hardship with kin seemed to be an important source of solace for Andrew's victims. There was a constant stream of visitors coming into the tent cities with supplies for relatives and friends, often bringing children to visit their parents. When asked who she felt had suffered the most, a young Hispanic mother living in a tent with sixteen other family members, replied, "I feel sorry for those women who are out there alone . . . having to go through it without any family. It's not that hard for me because I have my family with me."

HOMELESS FAMILIES

Estimates are that more than 180,000 people were homeless for some period of time after Hurricane Andrew (Governor's Disaster Planning and Response Review Committee 1993). It is officially estimated that about 100,000 people moved out of South Dade, at least temporarily (Smith and McCarty 1994). It was very common for families to split up, sending children or elderly members away from the terrible conditions of South Dade. When we asked about relocation on the Hurricane Andrew Survey, about 20 per cent of our total sample and 50 per cent of South Dade households reported that some or all members of their household had moved out of their home as a result of the storm.

Available temporary housing was grossly insufficient, even before the influx of emergency and construction workers. Most displaced households with resources – whether their own funds, insurance, or housing assistance vouchers – could not find a place to rent in South Dade. For lack of affordable alternatives, thousands were forced to remain in badly damaged homes or apartments, sometimes after being officially evicted or the property condemned. People made use of any available shelter; stories of families living in damaged cars and trucks, sometimes for weeks at a time, were common. A Red Cross caseworker told of a family of seven who were still living in one room behind their store 16 months after Andrew. Doubling-up was commonplace as thousands of displaced families moved in with relatives or friends, often into damaged homes. As a personal example, for nearly six months my elderly mother had two other house-holds living in her home (including co-editor Walt Peacock, his wife, and two children). Many families were forced to move several times during the long ordeal.

Staying with relatives

From our interviews with tent city residents (Chapter 6), it was clear that, in the absence of rental replacement housing, the preferred choice would have been to stay with relatives. The poignant quotation at the beginning of this chapter came from a Honduran woman living in a large tent with her husband, their two children, and three families of strangers.

Any relatives still in South Dade were likely to be living in badly damaged dwellings or even homeless themselves. In fact, about 38 per cent of our sample of tent city residents reported having relatives also living in the tents. One of my most vivid memories was an elderly African-American woman who talked while sitting on the side of her cot as she used a washboard and tub to launder her grandchildren's clothes. Living in the large tent with her were three adult daughters and their children – twelve family members in all – representing four destroyed homes.

About 20 per cent of the respondents in our Dade County survey had been forced to move out of their homes and, of these, about one-third had moved in with relatives. One of the most significant factors associated with having stayed with relatives was household income, as shown in Table 8.3. While a relative's home was the most frequent answer for all, it declined in importance with income. High-income households were about five times more likely to have rented temporary quarters.

While these findings suggest income differences, when other factors are controlled, the effect of income is less obvious. In the logistic regression model in Table 8.4 developed to explain the odds of staying with relatives, having resided in South Dade and sustained major damage are major factors. It is interesting

Table 8.3 Places of relocation of those forced to move by income

Where stayed	Under $20,000 % N=52	$20,000– $50,000 % N=90	Over $50,000 % N=77	Total % N=219
Relative's home	37.3	37.0	29.3	34.3
Apartment or trailer	22.9	22.2	23.8	22.9
Friend's home	16.2	22.4	12.8	17.5
Hotel/motel	5.8	9.5	28.5	15.3
Other	17.8	9.2	5.6	10.0

Source: FIU Hurricane Andrew Survey
Chi-square = 14.03, p ≤ .05

Table 8.4 Logistic regression models predicting relocation to relatives

	Stayed with relatives	*Relatives stayed with them*
Constant	−4.792	−3.864
Have relatives in area	1.134[a]	0.964[a]
	3.106	2.623
Black	−0.402	−0.175
	0.669	0.840
Hispanic	−0.168	0.145
	0.845	1.156
Income	0.075	0.113[b]
	1.078	1.120
Own home	−0.235	1.252[a]
	0.791	3.497
Insured	−0.393	−0.422
	0.675	0.656
Live in South Dade	0.529[c]	0.093
	1.698	1.098
Sustained major damage	0.892[a]	−0.033
	2.440	0.968
Chi Square	92.592[a]	46.418[a]

Source: FIU Hurricane Andrew Survey, n=1,068
Notes: a p≤.01
    b p≤.05
    c p≤.10

that being a South Dade resident is important, even after controlling for home damage. This is likely to be due to the tremendous difficulties associated with living in an area that sustained extensive damage to public facilities, businesses, and infrastructure.

Looking at it the other way, about 12 per cent of our total sample reported taking in relatives and the second model in Table 8.4 predicts the odds of that happening. Here, higher income and homeownership increased the odds. Among low-income households who had relatives move in with them, however, the chance of them still being there four months later was nearly three times higher than for the other two income groups. It would seem that, while poorer relatives are less apt to have kin move in, those who do are likely to stay longer.

GETTING AGENCY ASSISTANCE

In spite of the best intentions, the process for getting assistance from both public and private agencies tended to be filled with impediments. For many victims, just getting to the correct location to file an application was a major problem. Many cars had been destroyed by Andrew and no public transportation was available during those first few weeks. Because of the massive environmental destruction, including the loss of most street signs and familiar landmarks, locating agencies and assistance centers was problematic. Agencies had difficulty finding adequate facilities in which to set up their centers and many were forced to move several times. Women with primary responsibility for young children and/or without automobiles were at a distinct disadvantage in completing the arduous, time-consuming process of applying for help (Chapter 7).

The application procedures were often very complicated. For example, the first step to getting help from FEMA was to have been denied funds from the Small Business Administration (SBA) – not an easy concept for anyone to grasp. Many victims, especially those from destroyed mobile homes, had difficulty producing the required documentation to prove their pre-Andrew address, income, residency or citizenship status. It usually took several trips to the FEMA Disaster Assistance Centers (DACs) or other centers to complete various, rarely integrated, application processes. The sheer volume of needy families overwhelmed previously successful application procedures. FEMA attempted to relieve the pressure by instituting a toll-free dial-in service. However, this was of questionable value since most of South Dade was without regular telephone service for weeks, and applicants who waited in the long lines to use emergency telephones reported that the number was nearly always busy.

Successfully negotiating the aid process – getting all of the assistance for which a household qualified – typically took a great deal of time, energy, and skill in dealing with bureaucracies. These are assets in which poor families, in particular, are apt to be lacking. Victims who were adept at the process may have received more than they were due, while some needy families were under-served or not helped in a timely manner (cf. Due 1993a). As stated by the administrator of an agency coordinating home repair volunteers:

> There are families out there that, for whatever reason, didn't get into the system. Usually the less informed, [the less] educated people, or elderly. They can't get into a Red Cross center, a FEMA center. Or somebody told them not to even try . . . They have huge financial problems, but they also often have huge emotional problems, education problems, that compound their financial problems.

He gave an example of an elderly man whose daughter came to visit a year after the storm and found him living in a partially destroyed house with a tree through the roof. Yet, for some reason he hadn't asked for help. He had no FEMA registration number and the deadline had passed. Many agencies, including FEMA, eventually realized that the Hurricane Andrew situation called for new procedures reflecting a better understanding of the nature of this victim population, including more outreach and flexibility (*Miami Herald* 1993a, 10 May).

As a more viable type of temporary housing, FEMA eventually assigned trailers or mobile homes to many homeless families. While some were placed on private lots next to victims' damaged homes, most were located in cleared trailer parks. Eventually, more than 3,800 families were temporarily housed in FEMA trailers, but thousands more were denied them. While many were legitimately denied, others simply did not fit the guidelines that favored small nuclear family units with one head-of-household per address. FEMA and other agencies were ill-prepared to deal with South Florida's complex living arrangements. As expressed by another agency official:

> The rules needed to be relaxed because the families here don't fit the family pattern in the rest of the United States. You talk about extended families . . . the Haitians often have two, three families under the same roof. One may own the house, but all the rest live there. The same with Latino families. And that drove FEMA and the Red Cross crazy . . . They should have known from Los Angeles.

If multiple-family households wanted to stay together, no place would accept them. She continued, "You're going to need more than one voucher, and so, if you've got three families living under one roof, give a voucher to each family."

There were many ways in which recovery efforts were hampered by household definitions and expectations at odds with the realities of victims' lives. Miami's population tends to be dynamic, with new groups arriving all the time. New arrivals typically double up with family or friends already there, so that extended families and even non-related persons live together until they can afford separate residences. For example, about 2,000 Haitians had just been released from a refugee camp at the US military installation at Guatanamo, Cuba. Most were living with kin in South Dade. When their housing was destroyed, all the people living there were left homeless. Yet, according to FEMA guidelines, only one head of household per address could qualify for an Individual and Family Grant. Who actually received the money often depended on who happened to get to the DAC first to fill out an application, or in some cases, who was home when officials made their inspection visits. One caseworker working in the Haitian community made this observation:

> They were all living together . . . At first FEMA was saying everybody could go apply. So they would each go apply as head of a household. So what happened is, the ones from Guantanamo weren't working, but the ones who were really the head of that household, the ones who had signed the leases, weren't at home when the inspectors came . . . So all the ones from Guantanamo were receiving the checks because they were there. And they were the first ones to get rental money and trailers and everything, and those other ones who really lost everything did not get anything. And I had so many of those cases I had to call FEMA. They didn't agree with me. I said that this is not the head of household the way they thought.

About three or four months later they called asking me about the Haitian community. The ones stuck weren't getting anything. Then they started asking the others to return the rent money, etcetera. Of course, they couldn't . . . They had been living on it. So that's where it becomes hard. Now they're listening to us, but it's too late.

FEMA expected one application per address (by "the" head of household) and originally considered all duplications to be fraudulent. They were not prepared for the extensive number (40,000) which occurred. Much of the duplication resulted from the confusion of applications being taken many different ways, such as neighborhood "sweeps" and teleregistration, as well as unclear directions from workers. FEMA was also pressured by the Presidential Task Force to institute a "fast track" application process with an abbreviated form which did not list all household residents. Therefore, cross-checks could not be made at the time of the application. Of the 2,700 "fast track" applications filed, 1,200 turned out to be duplicates. In many cases checks were issued, necessitating the difficult and politically sensitive process of trying to recoup the funds later (Polny 1993: 33). Once they were aware of the magnitude of the problem, FEMA attempted to adjust its policies. Eventually FEMA, and later HUD, made special arrangements to accommodate some of the large and "non-traditional" families encountered in South Dade.

Another factor hampering family recovery was the reluctance of some land-lords to repair or rebuild rental housing. According to a government official interviewed six months after Andrew:

> In poor slum properties, such as in Goulds and Florida City, the landlord has collected the insurance, but hasn't bothered to fix the place up. And often the landlords are still collecting rent on those properties. Why do they [the renters] pay it? They're going to get thrown out anyway. The government may have condemned the property. Or these people have been evicted . . . people being thrown out of conditions that are already worse than Third World countries. There's no transitional housing now available. We're creating a whole new group of homeless. Children and families. It's a terrible, terrible thing.

National and international disaster response agencies such as the Salvation Army, American Red Cross, Mennonite Disaster Services, Lutheran Ministries, and United Methodist Disaster Response poured unprecedented resources into South Florida. Many, such as the Red Cross, engaged in long-term recovery projects and remained in South Florida for extended periods of time (Morrow, Peacock and Enarson 1994). These massive efforts, combined with the government programs and the contributions of hundreds of smaller organizations, resulted in the largest disaster response ever mounted in the United States.

An Unmet Needs Committee of public and private service providers was established by the Red Cross and later coordinated by Catholic Community

Services. It provided a very successful model for the collaboration of about forty agencies to address client needs not being met through traditional channels. As of March 1995, the group had served over 500 clients, cooperatively distributed in excess of $150,000 in donations from participating agencies, as well as another $137,000 from a Health and Rehabilitative Services grant. Neighbors Helping Neighbors, another emergent local group, coordinated volunteer efforts and contributions of thousands of people and groups. Through its efforts temporary housing was found for 500 families, donated goods estimated to be worth $250,000 were distributed, and many families were "adopted" by local churches and agencies to receive help.

SLOW RECOVERY

Recovery and rebuilding from Hurricane Andrew has been a very slow and uneven process. Nearly three years after, many families had yet to regain their pre-impact status. The destruction of entire communities placed extraordinary demands on insurance companies, government agencies, material suppliers, and skilled labor. Many homeowners lacked the necessary resources for repairing their homes, usually because they were uninsured, underinsured, or their insurance companies folded or paid too little (Chapter 9). Thousands without personal resources had their houses repaired by private voluntary agencies, many working through the Interfaith Coalition and the Salvation Army. Several groups, including Habitat for Humanity and ICARE, continue to be involved in long-term projects to build new affordable housing for low-income families. In July 1995, more than $10 million from the state Hurricane Andrew Recovery and Rebuilding Trust Fund was released for 1,000 homes still in need of rebuilding or repair, primarily because of contractor fraud or insurance company collapse (Musibay 1995).

FEMA mobile homes were initially assigned for six-month periods; however, when adequate housing was still unavailable, the deadline for moving out had to be extended several times. On the second anniversary of Andrew, more than 500 households remained in FEMA trailers (*Miami Herald* 1994, 24 August). A special task force of social service providers was formed to find housing for the last difficult-to-place cases, the majority of which were large multi-generational families, often headed by single mothers or grandmothers. The last FEMA trailer park was closed in February 1995, two and a half years after the hurricane. It took considerable effort to find housing for these remaining families. When no better solution could be found for the last few households, FEMA moved the trailers to a different park and provided them with three months' prepaid lot rental fees. After that, the families were on their own.

The reluctance to rebuild low-rent housing commonly encountered after disasters (Phillips 1993a) has also occurred in South Florida (Hartman 1993a). The official explanations for the slow rate of rebuilding of public projects and private Section 8 (subsidized) rental units, include problems with asbestos, funding, and contractors (Tanfani 1993). Whatever the reasons, few deny the reluctance to rebuild low-income multifamily housing. The housing shortage allowed landlords to raise rent prices and become more selective (*Miami Herald* 1993c, 7 November). For many families, especially large ones, finding afford-able housing became very difficult, if not impossible. In March 1995, Neighbors Helping Neighbors reported that 280 families still remained in need of housing assistance. New cases were still being brought before the Unmet Needs Committee, which had received a grant from United Way to continue its work.

EFFECTS ON FAMILY RELATIONSHIPS

The stress of life after Andrew has taken its toll on families. Numerous news-paper stories featured specific families in trouble (Swarns 1993). The most widely reported was a couple who "gave back" two adopted children because after losing their home they could no longer cope with the children's special problems (*Miami Herald* 1992k, 7 October). The number of parents relin-quishing custody of their children increased by 20 per cent in the first months after Andrew (Swarns 1993). Due to the destruction of many facilities for the elderly, there was an increase in applications for assistance in caring for frail elderly family members at home. Suicides were attributed to problems associated with the hurricane (Strouse 1995).

Divorce rates in Dade County were reported to be 30 per cent higher two months after the storm, suggesting an increase in marital breakups due to increased levels of stress (Swarns 1993). Examining this further, Figure 8.1 presents the number of divorce applications filed, along with other vital statistics, in Dade County from January 1990 until December 1994 (Office of Vital Statistics 1995; Keen 1995). Examination of this trend line suggests that divorce applications tend to peak early in the year and generally fall until they reach a low point in December. The year 1992 presents a somewhat different picture, with filings reaching a low in August, the time of the storm, and then beginning a general increase, peaking in the early months of 1993. The general increase during the last half of 1992, including a sharp rise from August to October, is quite different from normal yearly trends. However, the 1993 filings return to the former pattern. Overall, then, there does appear to be an alteration in filing patterns suggestive of increased marital disruption following Hurricane Andrew. However, it appears to be short-lived.

Marriages followed a similar pattern in the first few months after the storm – dropping and then rising again. Recorded births fell around the time of the hurricane, likely because many pregnant women left the area during that diffi-cult period. What is more interesting is that, contrary to a report in the local newspaper (Due 1993b), the number of births was down for April and May in 1993, about the time babies conceived after the storm would have been born. Recorded deaths initially dropped as well, likely also reflecting a population drop

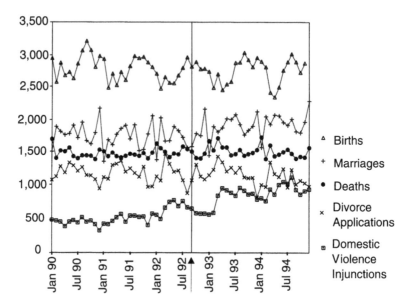

Figure 8.1 Monthly statistics for Dade County, Florida
Sources: Florida Vital Statistics; Dade Domestic Violence Unit
Note: Hurricane Andrew made landfall on 24 August 1992

as thousands relocated. Deaths were generally up in early 1993, reaching an all-time high seven months after the storm. The extent to which these changes can be attributed to the stresses of life after Andrew is speculative, but possible.

The pressures of coping in post-Andrew South Florida sometimes escalated into acts of violence. In a survey of 1,400 homes conducted for the Florida Department of Rehabilitative Services two months after Andrew, 35 per cent reported that someone in their home had recently been stressed to the point of losing verbal or physical control (Centers for Disease Control 1992). Alcohol-tax receipts indicated a significant increase in drinking (Wallace 1993), explained in part by the influx of construction workers and soldiers. Reports of domestic violence increased, prompting the addition of more staff at the crisis center in South Dade (personal interview with the Director of SafeSpace South). The switchboard of Miami's Helpline reported a 50 per cent increase in spousal abuse calls (Laudisio 1993). Related suits filed in Dade Circuit Court in 1992 increased by 98 per cent over the previous year (*Miami Herald* 1993b, 22 August). Again looking at Figure 8.1, after a decline right after Andrew – which the Court staff attributed to the inaccessibility of government services – the number of filings for injunctions rose sharply through the early months of 1993. This general tendency follows the upward trend which began several years ago and is probably attributable to reforms in the criminal justice system (Domestic Violence Coordination Unit 1995). One social worker had this to say about the clients he had been seeing from deep South Dade:

These people have a lot of inner strength. Very strong. I'm not sure how I would handle what they have had to. When you are in a crisis situation you become very resourceful. You have to. You're concentrating. The trouble is, it's like a crisis situation all the time. [You] can only deal with the present. The future is too overwhelming and depressing. The children have had a lot of problems because of the stress on their parents. Relocation, instability. Children become very anxious. They start acting out a little bit, and then mother starts acting out a little bit. We have worked with some of the families to try to relieve the stress and in a couple of cases we had to have HRS intervene because it was beyond [our resources]. We needed some protection for the children.

On our countywide telephone survey four months after Andrew, we asked respondents about stress in their household, as well as with relatives, neighbors, and friends. Table 8.5 presents, first for the entire sample and then for South Dade, the rates of increased stress reported for several relationships. Across Dade County, between 20 and 27 per cent felt that relations within their household (between partners, among adults, between adults and children, and among children) were more stressful after Andrew. The rates nearly doubled among households in hard-hit South Dade. While relations with relatives, neighbors, and friends were not quite as stressful, they were still nearly double for South Dade.

To understand better the factors associated with increased family stress, Table 8.6 presents a series of models predicting the odds of reporting higher levels of stress for four relationships. The predictors include: major home damage, living in South Dade, race/ethnicity (with Anglos again serving as the reference group), household income, homeownership, having minor children (under the age of 10), and being a single woman heading a household. As expected, living in South Dade and sustaining major home damage are important factors. In fact, the probability of experiencing more stress with partners doubles for respondents with major home damage and increases nearly three times for those in

Table 8.5 Perceptions of increased stress since Hurricane Andrew (percentages)

Respondents report more stress in relations:	Dade County N=1,318	South Dade County N=504
With their partner	27.6	56.1
Among adults in household	23.0	45.8
Between adults and children	21.6	46.6
Among children	20.9	43.0
With relatives	16.7	29.5
With neighbors	7.5	13.3
With friends	9.1	16.1

Source: FIU Hurricane Andrew Survey

159

Table 8.6 Logistic regression models predicting perceptions of increased stress in various relationships

	With partner N=757	With children N=495	With relatives N=961	With friends N=1,007
Constant	−1.202	−1.456	−1.416	−1.988
Black	−0.376[d]	−1.061[a]	−0.715[b]	−0.168
	0.687	0.346	0.489	0.845
Hispanic	−0.374[c]	−0.387	−0.535[a]	−0.864[a]
	0.688	0.679	0.585	0.421
Income	−0.008	−0.028	−0.034	0.023
	0.992	0.979	0.967	1.024
Minor children	0.174		−0.064	−0.524[b]
	1.190		0.938	0.592
Single mother	0.926[b]	1.002[a]	0.655[b]	0.685[b]
	2.524	2.724	1.924	1.984
Own home	0.079	−0.141	0.118	−0.144
	1.082	0.869	1.125	0.866
Live in South Dade	1.068[a]	1.419[a]	0.766[a]	0.605[c]
	2.908	4.132	2.152	1.832
Had major damage	0.720[a]	0.855[a]	−0.027	0.105
	2.055	2.351	0.974	1.110
Chi Square	74.112[a]	72.527[a]	31.774[a]	30.490[a]

Source: FIU Hurricane Andrew Survey
Notes: a $p \leq .01$
  b $p \leq .05$
  c $p \leq .10$
  d The first number refers to the unstandardized logistic regression coefficient and the second is its antilog.

South Dade. Of the two, living in South Dade has the most profound effect across all relationships, even after controlling for damage. Again, this is likely to be attributable to the difficulties associated with living amidst destruction and chaos. These data confirm what we observed first-hand – the families of Andrew have experienced a very difficult time.

While at first blush the results with respect to ethnicity and income may seem counter-intuitive, they are consistent with the literature. While minority and low-income households are impacted disproportionately, previous research indicates that families suffering high material losses, typically higher-income Whites, report high stress (cf. Dunal *et al.* 1985). Our results are consistent in that both Black and Hispanic households had lower probabilities of stress. The effect of income is in the negative direction, but having controlled for major damage, income differences are no longer significant. Being a single female head of household was associated with markedly higher stress levels across the board.

These women are 2.5 times more likely to report increased stress with their significant other, 2.7 times more likely with their children, and almost two times more likely to report higher stress with relatives and friends. It is generally apparent that the difficulties of dealing with the post-disaster situation resulted in high levels of family stress, particularly for these households.

WORKING CLASS HOMEOWNERS

Through several projects, including the tent cities interviews, Red Cross study, and Family Impact project, we gained insight into the plight of many of Andrew's poorest victims. During our time in the field we also became aware of visible neighborhood differences in recovery levels. As would be expected, wealthy and upper-middle class communities were the first to regain some normalcy. However, each time we passed through one particular community of modest single-family homes, we were struck by how little progress seemed to be occurring. Months after the storm, hundreds of badly damaged homes in South Miami Heights were virtually as Andrew had left them; yet, many appeared to be occupied. Despite broken windows, lack of electricity, and badly damaged roofs, at night we could hear voices and see kerosene lights inside many homes. Reconstruction was uneven – in any block there would be at least one or two badly damaged, unoccupied homes, several travel trailers or mobile homes sitting in front of damaged homes, occupied and unoccupied homes in various stages of repair, and a few looking brand new. According to Census records, this was a stable community primarily composed of minority working class home-owners.

We decided to study this community, and thus began a project during the summer of 1993 in which we completed door-to-door interviews with a randomly selected sample of over 200 households in South Miami Heights. About 95 per cent of our interviewees were buying their houses and the average residency was 10 years. The ethnicity of our sample was fairly representative of the community: 47 per cent non-Black Hispanic; 25 per cent Black, and 21 per cent Anglo. The median income was between $25–$30,000. Thus, the South Miami Heights survey provided considerable information about how families in a multicultural working class community of single-family homes were coping approximately one year after Andrew.

The conditions under which the interviews took place were difficult and emotionally draining. Interviews were completed in cramped travel trailers, rooms with no roofs, sitting on bare floors, standing outside while the respondent continued his/her repair work, and holding crying babies for stressed mothers. The noises of construction work and generators often made communication difficult – sometimes a car was the only quiet place to sit and carry out the interview. The interviewers often had to cope with erratic behavior and strong emotions as people vented their anger, stress, sadness, and fear. Many victims, both men and women, cried at some point during the interview.

Plate 8.1 Family life goes on in an unfinished home in South Miami Heights nearly two years after the hurricane
Source: Peggy Nolan

Plate 8.2 Reconstruction was uneven in neighborhoods like South Miami Heights, where some unrepaired houses still remained two years later
Source: Peggy Nolan

We asked respondents a series of questions about the effects of the storm on their property and family, their sources of assistance, and recovery progress. Over 70 per cent reported that their homes had been severely damaged or destroyed, with the median estimated repair cost category being $40–$50,000. When asked what percentage of their household possessions were lost, the median was an astonishing 72 per cent.

In spite of its devastating destruction, South Miami Heights had been virtually ignored by authorities during the early weeks after the storm. Public attention seemed to focus either on more well-to-do neighborhoods such as Country Walk or on poorer victims who had been living in trailers or low-rent housing in deep South Dade. We heard complaints such as, "We didn't exist after the hurricane. Homestead and Florida City got all the relief. We never saw police and soldiers. I called the television stations and demanded to know why." Or, "I don't think my area is on the map. Nobody ever came to my door to offer help. There were trees piled on my street for more than four months." One man told how neighbors got together and set up old Army tents on their lawns in the hope that anyone coming into the area to loot or cause trouble would think the military was there. Although complaints such as these were typical for unincorporated areas, this community seems to have been particularly neglected.

About one-third of our South Miami Heights sample had received some type of assistance from the American Red Cross, but almost all of the community's emergency and relief assistance, especially during the first weeks, was the result of the aggressive efforts of the local Catholic church, St. Joachim. Less than 18 per cent reported receiving assistance from FEMA – usually in the form of a small grant to help cover the loss of household goods, while about 7 per cent received SBA loans.

The damage was so extensive in two-thirds of these homes that some or all family members were forced to move out. The use of relatives' homes as places of refuge was higher among these working class homeowners than in our county-wide sample: 56 per cent stayed with relatives, 21 per cent resided in trailers on their lots, 18 per cent stayed with friends, and 9 per cent rented a place. Most who relocated were able to remain together as a unit. The average time of displacement was 95 days and at the time of the survey, almost a year after the disaster, 38 per cent of the households had still not returned entirely. Some families eventually returned to live in badly damaged homes for lack of viable alternatives. Clearly, the disruptions to these families were monumental.

Household stress was reported to be a large problem in about 42 per cent of the Anglo and Hispanic, and over 56 per cent of Black homes. More specifically, about 37 per cent reported more stress with their partners, 24 per cent with other household adults, 28 per cent with their children, 25 per cent among their children, and 20 per cent with non-household relatives. While these rates are lower than those reported on the Hurricane Andrew Survey, it should be noted that a year had now passed. Among those reporting increased parent–child

Plate 8.3 Destroyed neighborhoods were especially hard for families with small
children, who often had no safe place to play
Source: A. Enrique Valentin/*Miami Herald*

stress, 85 per cent had been relocated. Changes in finances were also strongly
correlated with increased household stress and were more apt to be a problem
in non-Anglo households.

We were interested in knowing more about the specific kinds of problems
these families were encountering. Respondents were given a list of areas, both
internal and external to the household, and asked how problematic they were.
The areas are rank ordered in Table 8.7 based on the percentage in the total
sample indicating each to be a major problem. Problems associated with living
in and trying to rebuild damaged homes ranked the highest. Dealing with
mortgage companies to get insurance money cleared, dealing with building
inspectors, and living in damaged homes were causing major problems to most
households, followed by problems related to neighborhood conditions, living
in temporary quarters, dealing with insurance companies, and dealing with
contractors.

There were some ethnic differences. While Black families were having less
problems associated with rebuilding, they were more likely to be troubled by living
in damaged homes or temporary quarters. They also appeared to be experiencing
more problems associated with community changes, such as neighborhood
conditions, transportation, and job relocation. Hispanic families were more
likely to report a problem associated with the loss of household members while

Table 8.7 Extent to which South Miami families perceive problems in selected areas by ethnicity (percentages)

Perceived to be a large problem	Anglos	Blacks	Hispanics	Total	Significance[b]
Dealing with mortgage companies to clear insurance money	67.5	48.6	67.7	63.5	a, c
Dealing with building inspectors	52.5	37.5	76.0	63.1	a
Living in damaged home	59.0	62.9	58.8	59.7	a
Neighborhood conditions	54.5	60.0	39.3	47.1	a
Living in temporary quarters	44.8	61.1	38.4	45.7	a
Dealing with insurance companies	33.3	26.3	48.5	40.4	a
Dealing with contractors	37.5	18.2	44.6	37.2	a
Unemployment	11.4	29.4	29.6	25.3	a
Household finances	13.6	39.5	19.8	22.3	a, c
Neighborhood crime	34.1	23.1	16.0	21.7	a
Transportation	2.4	28.2	16.7	15.9	a, c
Job relocation	6.5	20.7	16.7	15.1	
Dealing with agencies	11.1	20.0	12.5	14.5	
Behavioral problems with children	18.5	17.6	10.4	13.8	
Family violence	17.1	11.1	5.2	8.9	a
Gain of member(s)	14.3	0	3.6	4.8	a, c
Loss of member(s)	3.8	0	13.3	4.3	

Source: South Miami Heights Study, n=209
Notes: a p≤.05
　　　b Unless otherwise stated, significant difference between highest and lowest percentages
　　　c Significant difference between the most extreme percentage and the other two

for Anglos it was more likely to be the gain of new ones. Both minority groups had unemployment problems. While Black families were more likely to be having financial problems, over one-third of all South Miami Heights families reported a decrease in income since the hurricane. While not statistically significant, Black respondents were twice as likely to report problems in dealing with agencies. Anglos were more apt to mention neighborhood crime as a large problem. The Hispanic pattern was most similar to the Anglo, but even more likely to include problems in dealing with insurance companies, contractors, and mortgage companies.

Private insurance was the primary mode of post-disaster economic assistance, with 81 per cent of the families reported receiving no other rebuilding funds. Nearly 60 per cent did not believe their insurance settlements would be sufficient for repairing their homes. An important factor in recovery progress appeared to be the insurance company involved. Nearly all of the homes in which the repairs or rebuilding had been completed were insured by one of the major companies and had received early settlements when contractors, labor, and supplies were still available. (A more complete discussion of the insurance situation follows in Chapter 9.)

The interviewers encountered many homes in South Miami Heights where families were still suffering. This is supported by the data in Table 8.7 in which, for example, over 17 per cent of the Anglos reported the occurrence of household physical violence. About one-third of the parents had experienced increased behavior problems with their children since the storm. Looking more closely, the category of respondents reporting the highest stress levels was Anglo mothers with small children whose homes had sustained major damage. One mother of two children, ages two and five, admitted that at one point she had been on the brink of physically abusing her children and had been ready to leave her husband. She sought help through her church and was still in counseling at the time of the interview, and one of her children was on sedatives. Other respondents talked about bored and hyperactive children with too much time on their hands. Many South Miami Heights children were now attending a different elementary school on half-day shifts while theirs was being rebuilt. All neighborhood playgrounds were gone. The streets were full of unsafe construction debris. Even homes often lacked a safe place to play. It's no wonder children and their parents were having a hard time.

About 12 per cent of our sample indicated a need for counseling. As one woman described the problems her family was having, including violence between her husband and son, she said, with tears in her eyes, "It destroyed part of my life. It will never be the same again." Another man began crying as he said, "I can handle it all except the effect on my daughters." An adult daughter told about finally having to give up and move her mother to a nursing home in another county because she just couldn't care for her under the difficult circumstances. An elderly woman living with her disabled husband and twelve children and grandchildren in a crowded, damaged home kept saying, "I just can't get it together . . . I can't get it together."

Our interviews with care providers, as well as victims, supported other research (cf. Bates *et al.* 1963), which found that family roles change after a major disaster. Parents were having trouble meeting their children's needs and often found it hard to maintain control while living in temporary housing and sub-standard conditions. Children sometimes served as spokespersons for parents whose English was not good enough to deal with the bureaucracies. Men who took pride in their provider role had suffered from not being able to adequately protect and take care of their families. So much was beyond their control – job loss or change, long commutes to and from work, adjustors, inspectors, and construction workers who didn't show up on time, if at all. And, as discussed in the previous chapter, the family roles of women had expanded dramatically.

In contrast with what was occurring within their homes, most respondents reported no change in stress level in their relationships with non-household kin, neighbors, or friends. In fact, over one-third reported less stress with neighbors than before the hurricane. It was interesting that nearly 90 per cent felt the sense of sharing had been high in the neighborhood immediately after the storm and most said it was still high a year later. This supports the idea of a "therapeutic community" in the aftermath, and also indicates more lasting effects than usually reported (cf. Bates *et al.* 1963). In the absence of outside help, the residents of South Miami Heights were forced to rely on each other and perhaps that had contributed to improved neighborhood relations.

On the other hand, the unevenness of recovery progress often caused strain among neighbors. We listened to many stories about how "those people next door got more money." On some blocks there seemed to be a lot of resentment about neighbors who had received larger insurance payouts or FEMA grants. And, indeed, we visited some homes which had been totally remodeled, were filled with new furniture and appliances, and had new cars and boats on their driveways. It is easy to see how relative deprivation theory (Merton 1957) might be applicable here, causing those who had made less recovery progress to be resentful. While we found no evidence to support the assertions, some Anglos expressed strong opinions that their Black and Hispanic neighbors had gotten more government assistance. Many people voiced the opinion that the needs of the middle class had been ignored. Perhaps class and ethnic friction was already present in this community, but post-disaster frustrations had brought it nearer the surface. Erikson postulates that disasters "often seem to force open whatever fault lines once ran silently through the structure of the larger community, dividing it into fragments" (Erikson 1994: 236).

In addition to their own family's recovery, respondents often expressed serious concerns about the community. They perceived that a "lower class" was moving in, or that some homes would never be repaired and, as a result, their property values would be permanently lowered. They lamented the loss of playgrounds, stores, and other conveniences they feared would never be rebuilt. As one woman said, "We have lost our neighborhood."

Two years after we conducted the 1993 South Miami Heights interviews most homes in the community had been repaired or rebuilt. The rebuilt elementary school was better than the old one. Several social services agencies had permanently moved branch offices into the community. Habitat for Humanity had built thirty-one new homes for low-income families. Trees had been replanted along the roadways. Yet, there were visual reminders of the ordeal this community had been through – a scattering of homes still unrepaired, a few remaining trailers and broken fences. No doubt emotional scars remained within the homes as well.

UNEVEN OUTCOMES

Most of the families of Andrew have gone on with their lives, trying to put this experience behind them. Many homes have been improved in the reconstruction process – we often hear references such as "the new bathroom that Andrew built." Most replaced or restored public housing units are better than those previously there (Brennan 1993). Special projects have resulted in many victims becoming homeowners for the first time (Barry 1993) and several community-guided incentives to improve poor neighborhoods (Filkins 1993). The trailers for migrant farmworkers have been replaced with permanent housing and a community center has been built at the Centro Campesino Farmworker Center.

Under the terms of settlement of a lawsuit, FEMA agreed to allow approximately 8,000 victims of Hurricane Andrew who were denied benefits based on the head of household rule to reapply. FEMA also agreed to change the manner in which they will administer the individual disaster assistance program in future disasters (Coble 1995; Hancock 1993).

Many, if not most, communities have been improved by new roads, parks, landscaping, facilities, and services. Many families who permanently relocated are glad they were forced to move because they like their new communities better. Many men and women developed new skills and found new employment opportunities in the rebuilding efforts. In some cases, assistance monies were used by women to break away from abusive relationships.

On the other hand, it is unlikely that most families feel the suffering was worth it. Our work indicates important differences related to ethnicity and economic status in the ways in which the families of Andrew have been affected, but it has been painful for all. Anglo families with more resources expressed high levels of family disruption and stress. Yet, we know socioeconomic factors profoundly affect the extent to which families have the physical and human resources to withstand and recover from a disaster. In a very real sense, disasters rip the social fabric of communities, exposing their weaknesses and failures, as well as their strengths and successes. Hurricane Andrew provided ample confirmation of ways in which the poor get left behind. It may, however, be the case that working class homeowners are among those who suffer the most after a major disaster. While they have acquired some of the trappings associated with economic success, they may lack the "defense in depth" – the economic security, political and social influence, and personal power of the professional classes which can be especially crucial in times of crisis. And, if Hurricane Andrew is typical, working class neighborhoods are likely to be ignored by government, as well as private agencies, in the aftermath of a disaster.

CONCLUSIONS

After a disaster of this magnitude, a massive institutional response is essential for community and household recovery – and it needs to be planned to be long-term from the onset. However, it would be a mistake to neglect the importance of informal kinship links, particularly in families likely to be at a disadvantage in the highly competitive socio-political recovery process. The case of Hurricane Andrew provides evidence that urban families, regardless of ethnicity, tend to be embedded in kinship networks which can be a valuable resource when disaster strikes.

It can also be argued that most people did not receive help from their local kin. One reason for this may be that, due to the large area impacted, many relatives were also affected and, therefore, busy with their own problems. Other studies have found relatives to be an important source of assistance only if they live nearby and were not themselves affected by the disaster (Fogelman and Parenton 1956; Quarantelli 1960). Our findings also support the premise of Hill and Hansen, Bolin, and others (Hill and Hansen 1962; Bolin 1982) that families tend to be a less important source of assistance in large-scale societies. When other sources of assistance are available, reliance on kin diminishes, but remains important, especially when the need is acute.

The needs of South Florida families after Hurricane Andrew were on a larger scale than after any previous US disaster. Dozens of new and established agencies, both governmental and non-governmental, responded with unprecedented resources. Because the needs and clients were often qualitatively different, new approaches were called for, including more flexibility to meet the needs of multiple-family households and greater involvement in the long-term recovery of marginalized families. As a result, many agencies, such as FEMA and the ARC, eventually modified some of their policies and developed new strategies to fit the South Florida situation.

SOME POLICY CONSIDERATIONS

Policies which actively facilitate the optimum use of kin networks, such as grouping relatives together in temporary housing, should be adopted. Qualification criteria for assistance should not penalize multi-family households. Just because families were living "doubled up" before a disaster does not insure that they will be able to do so in the post-impact housing market. Policies must be developed which deal effectively with multiple head of household situations. Serious efforts are needed to find ways to protect the interests of women and their dependents better during the assistance process.

As recovery moved beyond cleanup activities to reconstruction, families relied less on kinship networks and increasingly on personal and institutional resources, such as private insurance, government programs, and the assistance of non-profit agencies. Having the ability to access these resources effectively was important to the negotiation of a timely and successful recovery. Clearly, every effort should be made to coordinate the various assistance programs effectively and to simplify the application processes. Many of the adaptations made throughout the Hurricane Andrew response were positive ones and should be developed into policies ready to be put into place at the onset of the next disaster response.

Natural disasters of this magnitude place a tremendous strain on families, particularly those with chronic problems. Families with strong internal bonds are better equipped to negotiate the difficult recovery process. Measures which strengthen these bonds, or at least do not weaken them, should be promoted.

169

An important part of institutional response, and one having a large payoff in terms of household recovery, is the provision of counseling and social services – such as crisis nurseries in highly impacted neighborhoods – to help families make it through the long recovery ordeal intact. New approaches to disaster response are needed which assertively address the affective needs of families, as well as the physical needs of households.

9

ETHNIC AND RACIAL INEQUALITIES IN HURRICANE DAMAGE AND INSURANCE SETTLEMENTS

Walter Gillis Peacock and Chris Girard

Look . . . look at that . . . No one escaped the impact of Hurricane Andrew.
Comment by public official viewing the damage

We don't need to worry about housing, it will take care of itself.
Opinion expressed at a redevelopment planning committee

These statements capture common beliefs regarding disaster impact and recovery. The first expresses the widely held view that natural disasters are impartial, indiscriminate forces that potentially impact everyone. This belief that disasters threaten everyone equally because they act in a quirky or haphazard fashion is reinforced by newspaper photographs of houses located next to each other in disaster-ravaged neighborhoods: one is destroyed while the other is relatively unscathed. In the face of such seemingly random outcomes fashioned by powerful natural forces, social factors that affect so much of our everyday lives – such as wealth, race, and ethnicity – are deemed to be of little or no import.

The exclamation that "No one escaped the impact of Hurricane Andrew" was spoken as the official's helicopter circled above a million-dollar residence where the barrel-tile roof had sustained damage to a corner section, the pool had turned dark green, and a surrounding grove of trees had been flattened. Not five minutes before, he had flown over acres of torn and twisted metal that once had been a mobile-home park: Not a single home was left standing. It looked like a very large scrap yard with none of the piles of metal recognizable as homes. While few of South Dade's households were spared the impact of Hurricane Andrew, it is hard to equate the total destruction of countless homes with one mansion's minor roof damage and stagnant pool. In this chapter, we will investigate the inequitable destruction of Hurricane Andrew and to what extent the differences were attributable to the nature of the community's ecological network.

The second statement quoted above, suggesting that "housing will take care of itself," was spoken during one of the many public/private sector meetings attempting to plan and coordinate recovery. It was made in an attempt to correct the flow of the meeting's conversation as it began to veer toward dealing with housing issues. The speaker indicated that the committee need not concern itself with housing because the market would rebuild and replenish the housing stock of South Dade. This outlook reflects the general logic that undergirds disaster policy in the United States: while government may have a role in disaster recovery, it will primarily be private enterprise exercised in the marketplace that will rebuild impacted communities.

Clearly, housing recovery depends on market mechanisms. Owners of damaged residential buildings (whether corporations or individuals) rely on insurance obtained in the private sector to pay for the services of profit-seeking contractors and builders. The literature on disaster recovery in the United States indicates that insurance is one of the most important determinants of recovery, particularly housing recovery (Bolin 1982; Drabek and Key 1982; Bolin and Bolton 1983; Bolin and Bolton 1986). With the infusion of insurance money, the construction industry expands rapidly and after a few years housing stocks tend to rebound to pre-impact levels (Wright et al. 1979; Friesema et al. 1979). However, it may not be the case that all households and neighborhoods fare equally well in the recovery process (Cochrane 1975).

The assumption that the market will rebuild housing may be overlooking important failures in market mechanisms. Even when a disaster is not present, the housing market in the United States has systematically failed to provide housing for many low-income households, many of whom are ethnic and racial minorities (Lake 1980; Bratt, Hartman and Meyerson 1986; Horton 1992; Alba and Logan 1992). In part, this is because low-cost housing is not the most profitable sector and because racial discrimination in the selling, financing, and renting of housing has yet to disappear, although it has come under legal attack and much has been done to combat it (Lake 1980; Guy, Pol and Ryker 1982; Sagalyn 1983). Further blocking access to housing is the neglected and more hidden problem of home financing – homeowners' insurance (Squires, Dewolfe and Dewolfe 1979; Badain 1980; Squires and Velez 1987; Squires, Velez and Taeuber 1991). Minorities are often unable to obtain a mortgage because they are unable to obtain the required property insurance. In light of these realities, the assertion that "housing will take care of itself" requires further scrutiny, especially given that pre-existing market failures are likely to play themselves out in the recovery processes.

This chapter will examine how ecological networks may produce racial and ethnic pockets of vulnerability to natural disasters – pockets that are potentially more susceptible to damage and which experience greater recovery obstacles. This focus emerges from a social science literature suggesting that poor and minority populations are impacted disproportionately by natural disaster agents, whether these agents appear in the form of hurricanes, tornados, earthquakes, or floods (Moore 1958; Moore, Bates and Parenton 1963; Peacock and Bates 1982; Bolin 1982; Bolin and Bolton 1986; Peacock, Killian and Bates 1987).

172

The literature has devoted less attention to racial and ethnic differentials with respect to insurance coverage and the impact of this on recovery. This chapter will address both of these issues by looking at damage and recovery in the area impacted by Hurricane Andrew – an area which had a good representation of Black, Hispanic, and Anglo homeowners.

DISASTER DAMAGE: WHY RACE AND ETHNICITY MATTERS

If disasters only struck homogeneous households residing in similar physical structures, it might be possible to suggest that the damage inflicted by a disaster agent is simply a function of the agent's magnitude and where it strikes. But in reality, the social units that comprise human communities are far from homogeneous. Our homes are not located in nondescript spacial areas, but rather in neighborhoods and, as any real estate broker will tell you, *location is everything*. Some neighborhoods are on the *wrong* side of the tracks, the *bad* side of town, or in slums and urban war zones. Others are on the *right* side of the tracks, *uptown, upscale*, or on the *good* side of town. Not only are the homes in which we reside and their locations imbued with socially significant meanings, but we cannot pick and choose from among neighborhoods that are socially evaluated as good or bad like fruit on a tree. In today's world – a world in which the commodity form has "metastasized" (Fjellman 1992: 4) – both our homes and their locations are commodities (Logan and Molotch 1987). To live in the best housing on the right side of town requires financial resources and, all too often, the right types of social characteristics. The urban landscape of the United States is one of segregated neighborhoods in which the distribution of quality housing is far from a random process.

Minorities, particularly Black households, are disproportionately located in poor quality housing segregated into low-valued neighborhoods. This segregation creates *communities of fate* (Stinchcombe 1965; Logan and Molotch 1987) that can take on added salience in a disaster context. Race and ethnicity are linked to housing quality – not because of ethnically based cultural variations in housing preferences as is true in some societies (Peacock and Bates 1982) – but because race and ethnicity are still important determinants of the economic resources, such as income and credit, critical for obtaining housing (Bolin and Bolton 1986). Thus, disadvantaged minority households are more likely to live in inadequate and poorer quality housing that may well be subject to greater damage. Moreover, race and ethnicity may be linked to housing quality because of segregation (Bolin and Bolton 1986).

The Miami metropolitan area was recognized as one of the most segregated urban areas for Blacks in the United States from 1940 to 1960 (Sorenson, Taueber and Hollingsworth, Jr. 1975). Though segregation has lessened somewhat today, as will be discussed further in the next chapter, over 45 per cent of Miami's Black population continues to live in blocks that are over 90 per cent Black (page 195).

Ultimately, we argue that the relationship between social ecology and disaster-generating natural forces is interactive (Blaikie *et al.* 1994). In advancing this argument vis-à-vis Hurricane Andrew, we acknowledge that the distribution of damage was principally a function of wind, particularly winds associated with the northern edge of the eye and the eye wall as it traversed Dade County. Analysis suggests that the damage was quite uneven, with pockets of extensive damage (Leen, Doig and Getter 1993; Fronstin and Holtmann 1994; Wakimoto and Black 1994). Although natural forces such as shearing played a role, social factors – such as race and ethnicity – cannot be ignored.

HURRICANE ANDREW DAMAGE

Our analysis of ethnic and racial differentials is based on the Hurricane Andrew Survey. The survey asked respondents if their homes were damaged and, if so, whether the damage was slight, moderate, or major. Respondents reporting damage were further asked about damage to windows and roofs: if their homes lost none, a small portion, about half, or most of the windows, and whether their roofs suffered no damage, damage to the covering (the tiles or shingles), damage beyond the covering, or major structural damage. Although the answers required an evaluation by the respondent, they allow for a comparison among households. Such a comparison would not be possible if we had asked about dollar amounts of damage, which are, in part, a function of the overall value of the property.

As described in previous chapters, the sample was divided into three racial/ethnic categories: Anglos, Blacks, and Hispanics. Respondents who self-identified as Hispanic were recorded as such, regardless of race. The remainder of respondents were divided into either Black or White (Anglo) categories. Each household, then, was ethnically classified on the basis of the respondent's self-identification. In part of our analysis, Hispanics were further subdivided into those with or without a Cuban background to produce a fourth ethnic/racial category. The latter distinction is important because Cubans have achieved economic and political prominence in Dade County (Chapter 3).

Damage levels by race and ethnicity are shown in Table 9.1. Although Blacks indicated higher levels of overall damage and roof damage, these differences are not statistically significant. Overall, while there are a few statistically significant results, the patterns are not simple nor consistent with regard to ethnicity. The problems with this very simple analysis are first, that other important factors such as location relative to the hurricane's path and type of structure are not controlled and second, each dimension of damage is considered separately which may not give us an adequate assessment of total house damage.

These shortcomings were readily overcome. First, the three damage measures were factor analyzed and then combined into a single index to give a better overall assessment of damage. The factor analysis generated a single factor, accounting for approximately 78 per cent of the observed variance in the three

Table 9.1 Race/ethnic differences in home damage from Hurricane Andrew

	Anglo % N=453	Black % N=191	Hispanic % N=624	Total % N=1,268	Chi-square[b]
Overall:					9.9
No damage	31.7	37.8	34.4	33.9	
Slight damage	30.9	29.7	35.7	33.1	
Moderate damage	21.5	18.8	18.9	19.8	
Major damage	15.9	13.7	11.0	13.1	
Lost windows:					9.6
None	68.1	69.4	73.2	70.8	
Small portion	16.2	17.6	17.2	16.9	
About half	12.6	3.4	3.3	3.8	
Most	15.8	9.6	6.3	8.4	
Roof damage:					17.9[a]
None	47.8	55.5	46.6	48.4	
Tiles/shingles	26.3	16.2	31.2	27.2	
Beyond covering	12.7	12.6	9.8	11.2	
Major structural	13.1	15.8	12.4	13.2	

	Anglo % N=453	Black % N=191	Non-Cuban Hispanic % N=245	Cuban % N=379	Total % N=1,268	Chi-square[b]
Overall:						14.7
No damage	31.7	37.8	37.9	32.2	33.9	
Slight damage	30.9	29.7	36.2	36.2	33.1	
Moderate damage	21.5	18.8	14.8	14.8	19.8	
Major damage	15.9	13.7	11.1	11.1	13.1	
Lost windows:						70.8
None	68.1	69.4	73.7	72.8	70.8	
Small portion	16.2	17.6	14.4	19.1	16.9	
About half	4.8	3.4	3.9	2.9	3.8	
Most	10.8	9.6	8.1	5.1	8.4	
Roof damage:						20.8[a]
None	47.8	55.5	50.6	44.0	48.4	
Tiles/shingles	26.3	16.2	30.0	32.0	27.2	
Beyond covering	12.7	12.6	8.2	10.8	11.2	
Major structural	13.1	15.8	11.2	13.2	13.2	

Source: FIU Hurricane Andrew Survey
Notes: a p≤.05
    b Chi-square tests based on 4 × 3 tables in upper panel and 4 × 4 in lower

measures.[1] The general damage measure had the highest factor loading (.9287), followed by roof damage (.8951), and window damage (.8302). Second, simple cross-tabulations were replaced with a regression analysis of damage, permitting several control variables to be entered. The control variables included were the household's proximity to the storm's path, whether the household was located in an evacuation zone, whether the household boarded its windows, household income level, and whether the household resided in a mobile home, apartment, duplex, or townhouse (compared to a single-family dwelling as the reference category). Annual household income served as a control for socioeconomic status, but was also expected to capture some variations in housing due to economic factors. A household's proximity to the hurricane eye was measured by locating it in one of five zones based on the resident's ZIP Code. As illustrated in Figure 9.1, the zones, moving from south to north, are: (1) the eye (county line to SW 200th Street), (2) the north eye wall (between SW 200 Street and SW 104th Street), (3) the adjacent edge (between SW 104th Street and SW 88th Street – Kendall Drive), and (4) north of edge (between Kendall Drive and Flagler Street). The remainder of Dade County is considered the fifth zone and serves as our reference category.

The first set of equations in Table 9.2 presents results for all households in Dade County, whereas the second set pertain only to homeowners. The first models in both sets show that, when controlling for other influences, Blacks and Hispanics sustained more hurricane damage than Anglos. The second equation in both sets indicates that the levels of damage reported by Cuban-Hispanic households were higher. Obviously, a household's location relative to the storm's path was among the most important factors determining damage. Note that the coefficients are positive and increasingly larger the closer one moves to the eye. Households located in evacuation zones also suffered significantly higher levels of damage than those outside evacuation zones, net of other factors. The type of residence was also critical: residences located in apartments and duplexes reported lower levels of damage than single-family dwellings, whereas mobile homes fared significantly worse.

Although these results are consistent with the general expectations from the literature – namely, that a disadvantaged minority group such as Blacks would be more vulnerable to hurricane damage, there is a twist. While it may be possible to suggest that Cuban-Hispanics are a minority at the national level, they are neither disadvantaged nor a minority in Dade County. And yet, their reported damage levels are higher than the reported damage levels for Anglos. Why were Blacks and Cubans seemingly more vulnerable to hurricane damage?

To try to answer this question, we took the analysis a step further by assessing whether ethnic segregation, in part, accounts for these ethnic differences. This was accomplished by including a segregation measure indicating whether a household was located in a postal ZIP Code area containing 75 per cent or more Black residents. A similar measure was used for ZIP Code areas with 75 per cent or more Hispanic residents. It must be noted that employing ZIP Codes in this

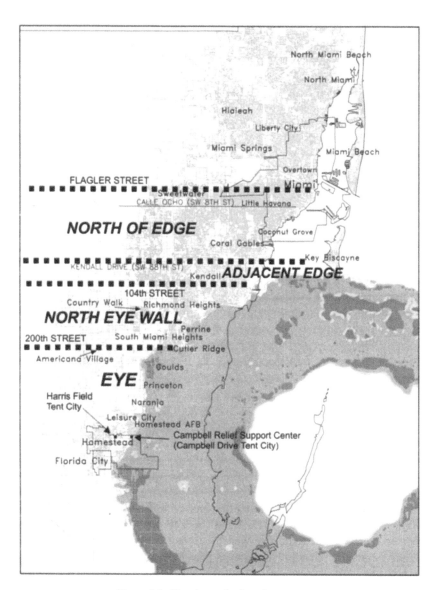

Figure 9.1 Hurricane Andrew impact zones
Source: Hugh Gladwin/IPOR

Table 9.2 OLS regression models predicting home damage

| | Total Dade sample | | | | Homeowners | | | |
| | Model one | | Model two | | Model one | | Model two | |
	b	*B*	*b*	*B*	*b*	*B*	*b*	*B*
Constant	−0.33[a]		−0.34[a]		−0.44[a]		−0.44[a]	
Black	0.16[a]	0.06	0.17[a]	0.06	0.21[a]	0.07	0.21[a]	0.07
Hispanic	0.11[a]	0.06			0.19[a]	0.06		
Cuban			0.17[a]	0.07			0.22[a]	0.10
Other			0.03	0.01			0.14	0.05
Income[c]	−0.03	0.01	−0.03	−0.01	−0.10	0.01	−0.11	−0.03
Boarded up	0.01	0.01	0.02	0.01	0.05	0.06	0.05	0.02
Apt./condo	−0.47[a]	−0.20	−0.45[a]	−0.20	−0.39[a]	−0.11	−0.39[a]	−0.11
Duplex	−0.33[a]	−0.08	−0.33[a]	−0.08	−0.13	−0.08	−0.15	−0.03
Townhouse	−0.10	−0.03	−0.10	−0.03	0.08	0.02	0.08	0.02
Mobile home	0.72[a]	0.20	0.74[a]	0.20	0.53[a]	0.09	0.53[a]	0.09
Eye	1.93[a]	0.38	1.95[a]	0.38	1.97[a]	0.41	1.97[a]	0.41
North eye wall	1.77[a]	0.44	1.77[a]	0.44	1.84[a]	0.51	1.84[a]	0.51
Adjacent edge	1.08[a]	0.27	1.08[a]	0.27	1.28[a]	0.35	1.28[a]	0.35
North of edge	0.37[a]	0.17	0.37[a]	0.16	0.46[a]	0.21	0.45[a]	0.21
Evacuation zone	0.11[b]	0.05	0.12[a]	0.06	0.19[a]	0.08	0.19[a]	0.08
R^2	.46[a]		.46[a]		.49[a]		.49[a]	
Adj–R^2	.45		.45		.48		.48	
N	986		986		690		690	

Source: FIU Hurricane Andrew Survey
Notes: a p≤.05
　　　b p≤.10
　　　c Actual coefficient equal to b × 10^{-2}

manner is less than optimal because they are established for administrative purposes by the post office and consequently do not necessarily reflect socially homogeneous neighborhoods. Furthermore, as will be discussed in Chapter 10 (page 196), Hispanics in general and Cubans in particular do not experience the degree of residential discrimination experienced by Blacks (Boswell 1992; Massey and Denton 1993). As a result, no ZIP Code area contained 75 per cent or more Cuban residents. Were it not for the very high proportion of Dade's population that is Hispanic, it may have even been difficult to have located one with 75 per cent or more Hispanics. Nevertheless, this procedure does allow the possibility of assessing the consequences of variability in housing and segregation.

For comparative purposes, Table 9.3 presents basic damage models, plus models including the segregation measures. Model two in each of the sets shows that households in the most segregated Black areas suffered significantly higher

Table 9.3 OLS regression models predicting home damage with segregation measures

	Total Dade sample				Homeowners			
	Model one		Model two		Model one		Model two	
	b	B	b	B	b	B	b	B
Constant	−0.33ᵃ		−0.34ᵃ		−0.44ᵃ		−0.44ᵃ	
Black	0.16ᵃ	0.06	0.09	0.03	0.21ᵃ	0.07	0.06	0.02
Hispanic	0.11ᵃ	0.06	0.11ᵃ	0.06	0.19ᵃ	0.06	0.18ᵃ	0.09
Income^c	−0.03	0.01	−0.02	0.00	−0.10	0.01	−0.09	−0.03
Boarded up	0.01	0.01	0.04	0.02	0.05	0.06	0.03	0.02
Apt./condo	−0.47ᵃ	−0.20	−0.40ᵃ	−0.18	−0.39ᵃ	−0.11	−0.37ᵃ	−0.11
Duplex	−0.33ᵃ	−0.08	−0.38ᵃ	−0.09	−0.13	−0.08	−0.14	−0.03
Townhouse	−0.10	−0.03	−0.12	−0.03	0.08	0.02	0.05	0.01
Mobile home	0.72ᵃ	0.20	0.54ᵃ	0.08	0.53ᵃ	0.09	0.55ᵃ	0.09
Eye	1.93ᵃ	0.38	2.05ᵃ	0.41	1.97ᵃ	0.41	2.02ᵃ	0.42
North eye wall	1.77ᵃ	0.44	1.88ᵃ	0.48	1.84ᵃ	0.51	1.88ᵃ	0.53
Adjacent edge	1.08ᵃ	0.27	1.15ᵃ	0.29	1.28ᵃ	0.35	1.31ᵃ	0.35
North of edge	0.37ᵃ	0.17	0.43	0.20	0.46ᵃ	0.21	0.48ᵃ	0.22
Evacuation zone	0.11ᵇ	0.05	0.09ᵃ	0.04	0.19ᵃ	0.08	0.15ᵃ	0.07
>75% Black			0.37ᵃ	0.09			0.36ᵃ	0.08
>75% Hisp.			0.04	0.02			−0.02	−0.01
R^2	.46ᵃ		.49ᵃ		.49ᵃ		.50ᵃ	
Adj-R^2	.45		.48		.48		.49	
N	986		967		690		679	

Source: FIU Hurricane Andrew Survey
Notes: a p≤.05
b p≤.10
c Actual coefficient equal to b × 10⁻²

levels of damage, net of other factors. Moreover, for both the total sample and for homeowners, differentials between Black households and Anglo households disappear when location in predominantly Black areas is entered into the equation. There is not a comparable effect for being located in a predominantly Hispanic area and, when controlling for these areas along with other factors, we still find that Hispanic households report higher levels of damage than Anglo households.

This analysis suggests, then, that a key factor in the Black–White differential may have been residential segregation, whereas this factor was not operating with respect to Cuban or other Hispanic households. These findings are consistent with the lower levels of segregation of Hispanics in Miami. However, the reason for higher reported damage levels for Cuban households relative to Anglo households remains a mystery and it is also difficult to say why there is a neighborhood effect for Blacks. It may be that segregated areas have lower levels of overall housing quality, but other factors may be operating as well. Perhaps features of the storm itself – such as wind shearing – were worse in areas where more Blacks and Cubans happened to live. Regardless, it is clear that there were ethnic differences in reported hurricane damage. Having set the context, in terms of damage, let us now turn our attention to the principal mechanism by which households attempt to overcome disaster damage – insurance settlements.

HOMEOWNERS' INSURANCE:
THE CONSEQUENCES OF RACE AND ETHNICITY

In the United States, the foundation of the recovery process is the private insurance industry. The insurance industry accepts individual risks and spreads them across policy holders. The exceptions are a result of state and federal policies that have removed certain hazards, such as floods and earthquakes, from normal homeowners' insurance coverage. For example, the National Flood Insurance Program was established by the federal government to cover flood-related losses. Federal disaster policies implicitly assume that private insurance will be the major mechanism for recovery because most government programs – such as low-interest Small Business Administration (SBA) loans – only come into play if the household is without insurance or is significantly under-insured.

The assumptions underlying government policy overlook significant gaps in a market-based, insurance-driven recovery process. Minority households – Black and Hispanic – are likely to have inadequate insurance (Bolin and Bolton 1986). Policyholders in these groups often lack supplemental flood or earthquake coverage or options that include temporary housing expenses. Some simply have insufficient coverage levels. Yet, it has also been found that federal programs designed to assist those with insufficient or inadequate insurance are less likely to be utilized by minority groups (Bolin and Bolton 1986; Bolin 1984). These problems are embedded in a long history of markets failing to respond to the needs of minority groups, such as the need for home mortgages (Squires, Dewolfe and Dewolfe 1979; Squires and Velez 1987; Squires, Velez and Taeuber 1991).

Traveling the neighborhoods in South Dade after Hurricane Andrew was a graphic lesson in racial differences in insurance coverage. Due to the absence of street signs and even house numbers it was very difficult to find specific addresses. To aid insurance adjustors many homeowners spray painted the name of their insurance companies, along with their addresses, on the outside walls of their homes. Driving through predominantly White neighborhoods, such as the famous Country Walk development, revealed a virtual *Who's Who* of insurance companies: State Farm, Allstate, and Prudential. In contrast, a drive through predominantly Black areas showed names that were far from familiar, such as Utah, Delta, Ocean Casualty, and Florida Fire and Casualty. At minimum, this anecdotal evidence suggests some degree of the market segmentation reflected in the dual economy literature (Averitt 1968; Beck, Tolbert and Horan 1978). The major insurance firms – representatives of the *core* of the United States insurance industry – are prominent in neighborhoods in which the majority of occupants are members of Dade's dominant ethnic groups, Anglos and Cubans. In Black neighborhoods, on the other hand, the names of smaller and less well-known firms – representatives of the insurance *periphery* or secondary market players – dominated. Unfortunately, many of these companies failed following the storm.

For households having homeowners' insurance, respondents in two different surveys – the FIU Hurricane Andrew Survey and the South Dade Population Impact Study – were asked if the household filed an insurance claim and whether they had received a settlement. The results are presented for the home-owners throughout Dade and those residing in the hardest hit area, South Dade. Before seeing how the anecdotal evidence concerning insurance companies compares to the findings of the Hurricane Andrew Survey, we will examine ethnic differences in insurance coverage and payout (see Table 9.4). To begin with, few Anglo or Cuban homeowners – 2.7 and 4.1 per cent respectively – indicated they were without homeowners' insurance, compared to 9.3 per cent of non-Cuban Hispanic homeowners and 8.7 per cent of Black homeowners. In South Dade, a somewhat similar pattern is found. Blacks were almost four times more likely than Anglos to say that they had no homeowners' insurance, although the percentages were too small (5 per cent of Blacks versus 1.3 per cent of Anglos) for the difference to be statistically significant. Apparently, few South Dade homeowners in general, only 2.3 per cent in our sample, were without homeowners' insurance (not surprising since mortgage companies require it).

More telling – and ultimately disturbing – findings concern the ethnic/racial differentials that emerged among insured homeowners who indicated they had filed a claim and had received a settlement. Among these homeowners, Table 9.4 shows that 16 per cent of Anglo households in Dade County reported that their insurance company was not offering enough to cover their rebuilding expenses, compared to more than double that percentage for non-Cuban Hispanics (41 per cent) and Blacks (38 per cent). As is the case for many such comparisons, the Cuban percentage (24.7 per cent) is more in line with that of Anglos. In South Dade, the pattern is similar, although the percentage of Black households reporting insufficient settlement climbs to almost 45 per cent – far higher than any other group.

What would account for such striking racial and ethnic differentials among insured homeowners? Is this finding simply an artifact of differences in income or damage? To test for this possibility, we used maximum likelihood estimators to predict the log-odds of receiving a sufficient insurance settlement, control-ling for income and damage level. The results in Table 9.5 indicate that, when controlling for race/ethnicity, income has no statistically significant effect. Although damage level does significantly diminish the likelihood of getting a sufficient settlement for all of Dade County (first two columns), in South

Table 9.4 Race/ethnic differences in insurance coverage and settlements[b]

	Anglo	Black	Hispanic	Total	Chi-square[b]
	%	%	%	%	
Total Dade:					
Not insured	2.7	8.7	5.8	4.9	8.4[a]
Sample size	(366)	(123)	(386)	(875)	
Cost not covered	16.1	38.1	29.4	24.4	16.1[a]
Sample size	(217)	(49)	(225)	(490)	
South Dade:					
Not insured	1.3	5.0	3.8	2.3	3.4
Sample size	(238)	(40)	(106)	(384)	
Cost not covered	21.2	44.8	26.3	24.8	7.7[a]
Sample size	(189)	(29)	(80)	(298)	

	Anglo	Black	Non-Cuban Hispanic	Cuban	Total	Chi-square[b]
	%	%	%	%	%	
Total Dade:						
Not insured	2.7	8.7	9.3	4.1	4.9	13.3[a]
Sample size	(366)	(123)	(126)	(260)	(875)	
Cost not covered	16.1	38.1	41.3	24.7	24.4	22.9[a]
Sample size	(217)	(49)	(64)	(161)	(490)	
South Dade:						
Not insured	1.3	5.0	3.5	4.1	2.3	3.4
Sample size	(238)	(40)	(57)	(49)	(384)	
Cost not covered	21.2	44.8	34.1	17.9	24.8	10.5[a]
Sample size	(189)	(29)	(41)	(39)	(298)	

Source: FIU Hurricane Andrew Survey
Notes: a p≤.05
     b Chi-square tests based on 2 × 3 tables in upper panel and 2 × 4 in lower

Dade (second two columns) there is no statistically significant effect. The results for South Dade are probably accounted for by the smaller sample and less variance in damage since few houses suffered no damage. In any event, in Dade County as a whole, as well as in South Dade, the odds of respondents indicating a sufficient insurance settlement are one-third to one-half as great for Blacks and non-Cuban Hispanics as for Anglos. The odds of Cuban-Hispanic households receiving sufficient settlements, on the other hand, were essentially the same as those of Anglos. These ethnic/racial differences remain highly statistically significant, even when controlling for income and damage.

Table 9.5 Logistic regression models predicting sufficient insurance settlements to cover reconstruction costs

	Total Dade Sample				South Dade			
	Model one		Model two		Model one		Model two	
	Coef.	exp(b)	Coef.	exp(b)	Coef.	exp(b)	Coef.	exp(b)
Constant	−1.79[a]		1.81[a]		1.54[a]		1.57[a]	
Black b	−1.04[a]	0.35	−1.05[a]	0.35	−0.99[a]	0.36	−1.01[a]	0.36
B	−0.13		−0.14		−0.13		−0.14	
Hispanic	−0.69[a]	0.50			−0.35[a]	0.71		
	0.14				−0.07			
Cuban			−0.43	0.65			0.02	1.02
			−0.09				0.00	
Other			−1.24[a]	0.29			−0.64	0.53
			0.19				−0.10	
Income[b]	0.19	1.00	0.13	1.00	0.35	1.00	0.30	1.00
	0.06		0.02		0.05		0.04	
Damage	−0.32[a]	0.73	−0.30[a]	0.74	−0.25	0.78	−0.25	0.78
	−0.12		−0.12		−0.09		−0.10	
Chi-square	17.26[a]		22.87[a]		10.17[a]		11.67[a]	
R_L^2	.037		.048		.034		.039	
R^2	.039		.055		.039		.047	
N	436		436		269		269	

Source: FIU Hurricane Andrew Survey
Notes: a p≤.05
   b Actual coefficient equal to b × 10^{-2}

If income and damage levels do not account for rather substantial racial/ethnic differentials in reported insurance settlements, what does? In the South Dade Population Impact Study, conducted approximately one year after the Hurricane Andrew Survey, we for the first time investigated the possibility that differences in the salience of insurance companies across neighborhoods – as suggested by our anecdotal evidence – may have been a contributing factor. We compare households insured by a "top-three" company – State Farm, Allstate, and Prudential, ranked by market share in Dade – against all others. The results, presented in Table 9.6, give strong support to this hypothesis. To begin with, the percentage of homeowners indicating that their insurance settlements were insufficient was three times greater (25.2 per cent versus 8.5 per cent) when their insurers were not among the top three companies. Second, among those home-owners not insured by a top-three company, slightly over half of Black homeowners said they did not receive a sufficient settlement, compared to about one-quarter of Hispanic homeowners and one-eighth of Anglo homeowners. The three-way ethnic/racial comparison shows a high level of statistical signifi-cance, although the Black–Anglo difference is the greatest: Black homeowners

Table 9.6 Households receiving insufficient insurance settlements by race/ethnic group and type of insurance company

	Anglo %	Black %	Hispanic %	Total %	Chi-square[a]
Total South Dade:					
Cost not covered	9.6	30.7	14.8	14.7	39.46[b]
Sample size	(464)	(147)	(245)	(856)	
Insured by top 3:					
Cost not covered	7.9	14.4	6.6	8.5	3.41
Sample size	(241)	(62)	(127)	(429)	
Not insured by top 3:					
Cost not covered	13.5	51.6	26.5	25.2	34.14[b]
Sample size	(150)	(63)	(127)	(289)	

Source: South Dade Population Study
Notes: a Chi-square tests based on 2 × 3 table
       b p≤.05

were almost four times more likely than Anglos to report insufficient settlements. On the other hand, *there are no statistically significant differences among racial/ ethnic groups for homeowners insured by a top-three company*, even though Black homeowners were twice as likely to indicate they had received an insufficient insurance settlement. Put quite simply, racial/ethnic differences are not significant because the level of payout was quite high for the top-three companies, regardless of race: over 90 per cent of all homeowners and 85 per cent of Black homeowners said they had received enough money to cover repair costs.

However, the question remains: to what degree are ostensible racial/ethnic -disparities in receiving sufficient insurance settlements a result of differences between the top three and other insurers when controlling for other relevant factors? Table 9.7 presents two models in which the log odds of receiving a sufficient insurance settlement are predicted by being Black or Hispanic. Control variables include the mean house value on the block (as a substitute for income), whether the housing unit suffered major damage, and, in the last equation only, whether the policy was issued by a top-three company. The results show that the odds of a sufficient settlement are tripled by a top-three company, but even when controlling for this effect, remain less than one-third for Blacks compared to Anglos. Although controlling for top-three insurers does not eliminate the gap between Blacks and Anglos, it does reduce the gap between Hispanics and Anglos to statistical insignificance. In other words, any differentials that appear between Anglos and Hispanics disappear after controlling for having a policy with a top-three company. However, the differential between Blacks and Anglos remains. Unfortunately, data from the South Dade Population Study do not allow us to distinguish between Cuban- and non-Cuban Hispanics.

In view of company-related differences in reported insurance settlements, are top-three companies less likely to insure certain ethnic or racial groups? Table

Table 9.7 Logistic regression models predicting sufficient insurance settlement

| | Model one | | Model two | |
	Coef.	exp(b)	Coef.	exp(b)
Constant	2.26[b]		1.27[b]	
Black b	−1.30[b]	0.27	−1.23[b]	0.29
B	−0.20		−0.19	
Hispanic	−0.45[b]	0.64	−0.28	0.78
	0.08		−0.05	
Mean house value[a]	0.02	1.00	−0.04[c]	1.00
	0.06		−0.10	
Major damage	−0.50[b]	0.61	−0.23	0.74
	−0.10		−0.05	
Top three company			1.21[b]	3.35
			0.24	
Chi-square		39.97[b]		66.94[b]
R_L^2		.057		.113
R^2		.058		.121
N		839		701

Source: South Dade Population Study
Notes: a Actual coefficient equal to $b \times 10^{-4}$
b $p \leq .05$
c $p \leq .10$

Table 9.8 Households insured by a top three company by race/ethnic group and proportion Black in Census block

	Anglo %	Black %	Hispanic %	Total %	Chi-square[a]
Households insured by					
top three	61.9	49.1	64.5	60.5	8.38[b]
Sample size	(406)	(124)	(245)	(750)	

| | Proportion Black in Census block | | | | |
	≤25%	26–75%	>75%	Total	Chi-square
Households insured by					
top three	63.2	58.7	36.7	60.7	17.1[b]
Sample size	(645)	(54)	(64)	(762)	

Source: South Dade Population Study
Notes: a Chi-square tests based on 2 × 3 table
b $p \leq .05$

9.8 shows that over 60 per cent of Hispanics and Anglos were insured by a top-three company, whereas only 49 per cent of Blacks were. The difference is statistically significant. More striking are the apparent effects of Black residential segregation: in blocks with three-quarters or more Blacks (as indicated by Census data), the proportion of households (regardless of race or ethnicity) having a policy with a top-three company was 38 per cent, compared with 63 per cent of those in blocks with one-quarter or fewer Blacks. These findings are certainly suggestive of red-lining (refusing policies to residents of Black neighborhoods) by major companies.

These results are reinforced by regressions in Table 9.9 predicting the log-odds of a household having an insurance policy with a top-three company, controlling for the mean house value on the Census block (a proxy for income), and whether the home's value is in the top quartile (since the top-three firms generally do not target the upper end of the housing market). The first model shows significant racial/ethnic differences: the odds of having a policy with a top-three firm for a Black household are roughly half that of an Anglo house-

Table 9.9 Logistic regression models predicting if household was insured with a top three company

	Model one		Model two	
	Coef.	exp(b)	Coef.	exp(b)
Constant	0.46[b]		0.79[b]	
Black b	−0.59[b]	0.55	0.06	1.06
B	−0.11		0.01	
Hispanic	0.03	1.03	0.08	1.09
	0.01		0.02	
Mean house value[a]	0.02	1.00	−0.01	1.00
	0.06		−0.03	
Upper 25%	−0.61[b]	0.54	−0.64[b]	0.53
	−0.13		−0.14	
% Black			−0.01[b]	0.99
			0.19	
				1.00
% Hispanic			−0.00	
			−0.04	
Chi-square		14.03[b]		23.98[b]
R_L^2		.014		.024
R^2		.019		.033
N		735		735

Source: FIU Hurricane Andrew Survey
Notes: a Actual coefficient equal to b × 10⁻⁴
     b p≤.05

hold. In contrast, the odds of being covered by a top-three insurer are roughly the same for Hispanic households and Anglo households. The second model shows that, after controlling for the per cent Black on the block, the racial status of the household is no longer significant. Whereas the per cent Hispanic on the block has no effect, the per cent Black is statistically significant. These results reinforce the conclusion that residential segregation deters Blacks, but not Hispanics, from obtaining a policy from a top insurance company. In attempting to explain this ethnic difference, one cannot ignore the red-lining practices that have historically guided the banking and insurance industries.

These findings provide substance to the message already conveyed by the "handwriting on the wall" – the spray painted anecdotal evidence of insurance red-lining that was so visible throughout South Dade. The sum total of the evidence suggests that not only did Black households evince higher vulnerability to damage, they were also at much higher risk of failing to recover due to insufficient insurance settlements.

SUCCESSES AND FAILURES OF MARKET-BASED HOUSEHOLD RECOVERY

This chapter began with two major misconceptions concerning disaster impact and recovery. The first misconception is that disaster agents are basically indiscriminate and uneven damage can be attributed to quirks in the agent itself. In contrast to this assumption, we have attempted to show that the urban social landscape has pockets of vulnerability to disaster – pockets that are not only more susceptible to damage, but less resilient in the recovery process. In particular, our analysis has focused on the vulnerability created by residential segregation in conjunction with the possible effects of insurance red-lining. A second misconception – at least in part – is that housing will take care of itself in the recovery process. The underlying assumption after Hurricane Andrew was that the market would rebuild the housing stock of Dade County, and consequently, there was no need to focus concerted civic effort on housing recovery. However, the mechanism central to the market-based rebuilding process – homeowners' insurance – failed to provide sufficient funds for recovery for 15 per cent of the insured respondents in our survey. These 15 per cent were not randomly scattered across the entire population, but were disproportionately Black households, the same households that were more vulnerable to damage and were likely to be located in segregated neighborhoods. Again, our literature review suggests that these inequities may stem, in part, from red-lining problems experienced by Black households. It was also the case that Black households were at least three times as likely not to have homeowners' insurance as White households. Thus, inequalities within the ecological network that existed prior to Hurricane Andrew were likely to be exacerbated by the inequalities inherit in the market-based recovery process itself.

Although we have focused on pockets of vulnerability that tend to be passed over by market mechanisms, it is also true that the market did have spectacular successes in propelling the recovery process in Dade County. First, the vast majority of homeowners in the Hurricane Andrew Survey had insurance. Only 5 per cent of all Dade County homeowners reported not having insurance and that number fell to only 2.3 per cent in South Dade, the area hardest hit by Hurricane Andrew. Second, the majority of households reported insurance settlements sufficient to meet reconstruction and rebuilding costs. While approximately 76 per cent of respondents to the Hurricane Andrew Survey reported sufficient settlements four months after Andrew, that percentage increased to approximately 85 per cent, according to results from the South Dade Population Study conducted a year after the storm. Furthermore, it is important to note that Hispanics in general, and Cuban-Hispanics in particular, while minority groups at the national level did not seem to experience difficulties with insurance or insurance settlements when compared to Anglos. In fact, Hispanics were more likely than Anglos to have a policy with a top-three insurance firm, one of the most important determinants of receiving a sufficient settlement. These findings are likely to be attributable to the remarkable success of Cuban-Hispanics and the Cuban ethnic enclave in Dade County.

Marginalization of the already marginalized

Although we are not suggesting a conspiracy against Black homeowners, a convergence of obstacles results in a lack of adequate protection against life's disasters, both natural and social. Black households seeking the "American dream" of home ownership and financial security face a daunting task (cf. Logan and Molotch 1987). In order to find a home, they confront discriminatory practices from real-estate agents who steer them to predominantly Black neigh-borhoods (Pearce 1979). Blacks face additional discrimination when attempting to obtain a mortage. Not only are Black applicants more likely to be refused, the mortgage company is likely to require a more substantial down payment, in part because the property is not as likely to appreciate over time (Sagalyn 1983). A further impediment is confronted when attempting to procure homeowners' insurance. Due to problems of agency accessibility and traditional red-lining practices, their policies are more likely to be purchased from a company at the periphery of the insurance industry.

And then a hurricane ravages the neighborhood. Recovery is especially problematic in Black neighborhoods where poorer quality building construction may have made the neighborhood more vulnerable. Heavy damage is now coupled with insufficient insurance and inadequate settlements for rebuilding. Added to these is the fact that many of the companies that provided policies in Black neighborhoods in Dade County were the first to fail, thereby beginning the long and difficult task of obtaining funds from the Florida Insurance Guaranty Association, which "guaranteed" many policies of the failed firms.

Due to a combination of these factors, rebuilding in Black neighborhoods is likely to be much slower. Household recovery will be very uneven, with beautifully rebuilt homes standing next to condemned and abandoned properties. Moreover, looking at the bigger picture, we find that neighborhoods already at considerable risk are further marginalized in the post-disaster recovery process. This occurs not because of a single event, nor is it due to the disaster agent itself. Rather, it is *a series of obstacles built into the urban social structure* that places certain neighborhoods and households at substantially higher risk. The acute problems posed by Hurricane Andrew have only exposed the chronic problems already present. It is these chronic problems that must be addressed if pockets of vulnerability within the community's ecological network are to be fortified for the next disaster.

SOME POLICY CONSIDERATIONS

In light of the importance of insurance in a market-based recovery, government policies and programs, particularly at the state level, need to address the problems through a variety of ways.

- Public education and awareness programs. Funding should be made available to distribute information concerning insurance policies, rates, and consumers' rights regarding insurance. This information should be distributed through community development corporations, organizations specializing in low-income and minority housing, organizations specializing in post-disaster recovery, and other organizations that deal with housing and mortgages. Also, the development of standardized policies would greatly facilitate efforts to inform the public so that consumers can make intelligent choices.
- Minority agency placement programs and minority agent development programs. The private sector should be encouraged to develop programs that foster the placement of insurance agencies and agents in minority communities.
- Community Insurance Act legislation. This legislation would be similar to a Community Reinvestment Act that requires lenders to be responsive to the credit needs in their service area. However, this legislation would focus on insurance needs in particular service areas.
- Anti-red-lining and related policies. These programs would identify potential problems in red-lining and provide an objective evaluation of risk assumptions. Here are a few suggestions for gathering relevant information:
  - Require insurance companies to file quarterly disclosure reports which include the number and types of policies written, in effect, and canceled and, for each policy, the property's geographic location (plat description, tax assessment identification number, geo-code, and so on). If it is necessary to aggregate these data, the unit aggregation must be substantially smaller than the ZIP Code area, such as the Census block. Data on premiums collected and direct losses incurred should also be included.

189

- Require agencies/agents to file quarterly disclosure reports which would include their locations, underwriting activities, ethnic/racial characteristics of agents, and so on.
- Institute state insurance research programs. Using some combination of state and private funding, channeled through the state's appropriate insurance regulating entity, various research projects and testing programs concerning insurance availability would be conducted.

NOTES

1 The factor-scoring coefficients were as follows: overall damage, .39469; roof damage, .38040; and window damage, .35283. The factor was simply the linear combination of each standardized variable multiplied by its factor-scoring coefficient (Harmon 1967).

10

ETHNICITY AND SEGREGATION

Post-hurricane relocation

Chris Girard and Walter Gillis Peacock

Hurricane Andrew devastated a substantial portion of metropolitan Dade County. Barely missing downtown Miami, the storm's most powerful winds left a band of destruction 300 blocks wide. This band stretched across a rapidly developing, mostly residential part of the county we refer to as South Dade. Fully nine out of ten homes were damaged in this vast area, extending all the way from Kendall Drive in the north to the Everglades and Biscayne Bay in the south (based on the South Dade Population Survey). Altogether 49,000 homes were rendered uninhabitable (*Miami Herald* 1994, 24 August).

Because of hurricane damage, more than 180,000 residents left their homes for some period of time (FIU Hurricane Andrew Survey; also, see Smith and McCarty 1994). Official estimates indicated an immediate exodus of approximately 100,000 people from hurricane-ravaged South Dade (Metro Dade Planning Department 1993). Those leaving comprised almost 28 per cent of South Dade's 1992 estimated population (Smith and McCarty 1994), with most relocating north of the impact area.

South Dade's population hemorrhage was generated by harsh conditions in the areas hard-hit by the hurricane. In fact, many South Dade families were stranded in refuse-strewn, malodorous neighborhoods without electricity, running water, or any semblance of security for days, weeks, and even months. Throughout South Dade, a curfew was established to curb vandalism, burglary, and looting. Much of the area lacked street signs, traffic lights, shops, grocery stores and schools. Ultimately, access to jobs and services required residents to endure hours-long trips in heavy traffic.

Yet nearly half of South Dade's residents did not relocate after the hurricane. Given the harsh conditions, one may question why so many people stayed. The purpose of this chapter is to examine the factors that either inhibited or facilitated household relocation, be it temporary or more permanent. Did some people stay simply because their homes sustained less damage than those who left or did some residents experience barriers to leaving? One obvious barrier is insufficient economic resources. Finding substitute housing in an undamaged

191

Table 10.1 Racial/ethnic composition of South Dade, 31 July 1993

Zones[a]	% within ethnic group			% within each zone			Absolute change in per cent from 1990[b]		
	Black	Hisp.	Anglo	Black	Hisp.	Anglo	Black	Hisp.	Anglo[c]
1 Eye	56.88	40.19	32.07	30.28	31.19	38.53	7.71	0.56	-8.28
2A Lower north eye wall	34.90	21.72	28.79	33.60	25.65	40.75	6.81	0.46	-7.27
2B Upper north eye wall	5.69	13.84	17.39	11.52	36.11	52.38	5.12	6.64	-11.75
3 North of edge	2.54	24.25	21.76	3.84	47.18	48.98	-0.87	6.11	-5.24
Total	100.00	100.00	100.00	22.54	34.10	43.36	4.44	3.47	-7.91

Source: South Dade Population Study
Notes: N=1,508 housing units with 4,884 permanent residents; 39 additional households were interviewed for which race was not identified. It should be noted that the respondent's racial identification was assigned to the entire household.
a Zone 1: SW 200 Street and below; 2A: between 200 and 136 Streets; 2B: between 136 and 88 Streets; 3: between 104 and 88 Streets.
b Based on 1990 Census block data, 4,763 blocks south of 88 Street (Kendall).
c Non-Black, non-Hispanics may include a small percentage of non-Anglos.

area required adequate personal finances or insurance policies covering temporary living expenses. Some households were undoubtedly lacking these economic resources. In general, we know that households with fewer economic resources have more difficulty adjusting to disaster impact and recovering (Fried 1966; Cochrane 1975; Bolin and Bolton 1986; Morrow-Jones and Morrow-Jones 1991).

Race is another factor, perhaps less obvious than income, that may have affected relocation from South Dade's most damaged neighborhoods. Approximately 90 per cent of South Dade's Blacks live in the area (Zones 1 and 2 in Figure 9.1, page 177) with the greatest destruction and highest post-hurricane vacancy rates compared to the rest of South Dade. A substantially lower proportion of South Dade's Hispanics and Anglos (61 per cent of each) are located in this heavily damaged area. Based on these percentages alone, one would predict greater Black population loss in South Dade than Anglo or Hispanic. And yet, data from our South Dade Population Impact Study presented in Table 10.1 indicate that when comparing 1990 US Census data to our 1993 findings, the Black percentage in South Dade rose by nearly 4.5 points compared to a 3.5 percentage point increase in the Hispanic proportion, while the Anglo proportion decreased by nearly 8 percentage points. Blacks registered the highest increases below 136th Street (Zones 1 and 2A) – where the eye crossed and the greatest damage occurred.[1] To what degree is this increase in Black proportion a result of population trends underway before the hurricane and to what degree is this a consequence of barriers to Black flight from damaged neighborhoods?[2]

ETHNIC/RACIAL VARIATIONS IN RELOCATION

To understand better ethnic relocation patterns, we use three different surveys that include respondents in South Dade. These surveys, discussed in detail in the Appendix, have the advantage of being conducted at different intervals after Hurricane Andrew – specifically, four months (the FIU Hurricane Andrew Survey), one year (the South Dade Population Impact Study), and three years (the Homestead Housing Needs and Demographic Study). In each of these studies, respondents were asked about relocation after the hurricane. In some of the surveys respondents were asked why they left their homes and how long they were gone.

Results for these three projects, presented in Table 10.2, show that Anglo households were, in general, most likely to have left their homes, even after controlling for the level of damage. Generally, Blacks were the least likely to leave and Hispanics displayed an intermediate tendency. The differences between Blacks and non-Blacks (Hispanics and Anglos combined) are statistically significant in the South Dade and Homestead studies and the differences are largest when the period of relocation was three months or more. The results are not statistically significant in the Hurricane Andrew Survey, perhaps due to the smaller sample.

193

Table 10.2 Post-hurricane relocation by race/ethnic group

	Percentage leaving home because of Hurricane Andrew				Significance[a] difference	
	Anglo	Black	Hispanic	N	Black vs. other	Hisp. vs. other
Hurricane Andrew Survey 12/92:						
All of South Dade	54	53	50	488	–	–
Zones 1 and 2	64	57	63	294	–	–
Unit had major damage	81	70	76	237	–	–
Unit had major damage and left South Dade	36	26	49	179	–	–
South Dade Population Study 7/93:						
All of South Dade	38	28	34	1,173	b	–
Zones 1 and 2	34	27	41	599	c	c
Unit had major damage	58	42	45	582	c	–
Unit had major damage and left for 3 months or more	46	22	25	573	b	b
Homestead Housing Study 12/94:						
Left for any period of time	65	54	62	808	c	–
Left for 3 months or more	31	18	25	734	b	–

Notes: a Based on chi-square test for 2×2 tables cross-tabulating ethnicity (the ethnic group indicated versus the other two groups combined) by whether the household left due to hurricane damage.

b $p \leq .01$

c $p \leq .05$

What explains the tendency of Blacks to remain in damaged homes and neighborhoods? Perhaps low income or relatively tolerable levels of damage explain this tendency. Another possibility is racial discrimination in the housing market. To understand this latter factor, it is helpful to examine the history of residential segregation in Dade County.

Miami's history of Black segregation

In the 1920s, the majority of Miami's 25,000 Black residents lived in the area known as Overtown, in about fifty crowded blocks close to the central business district (Mohl 1991). In addition, some Bahamian-Blacks settled in a segregated section of Coconut Grove and there were several small Black farming communities along the Florida East Coast railroad tracks in South Dade (Kerr 1984; Allman 1987; Mohl 1991). In 1937, Miami's civic elite inaugurated a Black public housing project named Liberty Square, which soon spawned a burgeoning Black area now called Liberty City. When Liberty City reached its capacity, Blacks moved into adjacent White working class areas, especially to the north. This movement was met with protests and violent resistance, including Ku Klux

Klan cross burnings, bombings, and shootings (Mohl 1991: 124). To the present day, a process of succession-invasion, in which Whites move out as Blacks move in, continues in the northern parts of Dade County.

Although Blacks were no longer concentrated in a single core by the 1960s, 96 per cent lived within ten scattered, compact pockets throughout Dade County (Boswell 1993); see also Figure 3.2, page 45). This concentration subsequently subsided somewhat and yet, in 1980, 71 per cent of Dade's Blacks remained in these concentrated areas – the largest being Overtown, Liberty City, Carol City, and Opa-Locka. Several of the areas with Black concentrations in South Dade – Perrine, Goulds, Cutler Ridge, Homestead, Florida City, and Richmond Heights – were heavily ravaged by Hurricane Andrew (see Figure 9.1, page 177).

The high degree of segregation of Miami's Blacks, which has been discussed in terms of pockets or ghetto areas, can be appreciated more fully when measured by dissimilarity scores. These scores give the per cent of one ethnic group that would have to move to other areas, Census tracks for example, in order to achieve racial balance. In other words, a dissimilarity score of 50 indicates that 50 per cent of a specific ethnic group would have to move in order to have racial balance.[3] By convention, dissimilarity scores above 30 indicate moderate segregation and above 60 indicate high segregation (Massey and Denton 1993).

During the period from 1940 to 1960, the dissimilarity index for Miami approached 98, earning Miami the reputation of being one of the most segregated cities in the United States (Sorenson, Taueber and Hollingsworth, Jr. 1975). Currently segregation is still high, although its magnitude is typical for large metropolitan areas in the US (Massey and Denton 1993; Boswell 1993). Calculations based on 1990 Census block data indicate that 78 per cent of Dade County's Blacks would have to move to achieve racial balance, and fully 45 per cent of Blacks live in blocks that are 90 per cent or more Black. South Dade's level of segregation is somewhat lower, with an index of 64. If Census tract data rather than block data are used, the contrast between Dade and South Dade is more marked: 71 in Dade versus 52 in South Dade.

Studies have identified several mechanisms that maintain residential segregation. For example, there is ample evidence of *racial steering* of Blacks away from White neighborhoods by real estate brokers (Yinger 1991; Pearce 1979). This forces Blacks looking for properties in White areas to do so without professional assistance. Limited time and greater dependence on public transportation can also curtail the search for housing by minorities and the poor (Cronin 1982; Cadwallader 1992). Furthermore, the resulting ghettoization often results in a lack of knowledge, skill, resources, institutions, and networks that facilitate moving out of segregated areas (Wilson 1987; Logan and Molotch 1987). Finally, even after finding a suitable residence, Blacks confront discrimination when procuring financing and insurance, as well as prejudicial treatment by owners and neighbors in White areas (Squires, Dewolfe and Dewolfe 1979;

Badain 1980; Lake 1980; Leahy 1985; Bratt, Hartman and Meyerson 1986; Squires and Velez 1987; Horton 1992; Alba and Logan 1992). Given these barriers, many Blacks remain in predominately Black neighborhoods.

Trends toward Hispanic integration

Like Blacks, Hispanics in Dade County are concentrated in certain residential areas. However, the degree of concentration is moderate and would appear to stem more from continuing Hispanic immigration than from discrimination. During the period between 1960 and 1990, Hispanics grew from 5 to 49 per cent of Dade's population (Boswell 1993: 83). Newly arrived immigrants tend to settle in heavily Hispanic areas where they are fully accommodated within a Spanish-speaking enclave, regardless of English skills. Boswell (1992) describes these areas as "two wedges" – the first fans out west of Miami's central business district into Dade's northwestern quadrant containing Hialeah, and the second fans out from Little Havana into Westchester and Sweetwater (see Figure 3.3, page 46). Movement into these Hispanic residential areas has been largely voluntary because they provide immigrants unfamiliar with English a haven in which to establish a foothold (Portes and Rumbaut 1990). Once established, however, Hispanics have tended to relocate throughout Dade, resulting in only moderate levels of segregation (Portes and Stepick 1993). Indeed, the Hispanic experience, particularly for Cubans, can be generalized as rapid economic advancement, political empowerment, prominent roles in the real-estate, construction and FIRE (financing, insurance, real-estate) sectors and, as a result, residential dispersion throughout the Miami area (Allman 1987; Portes and Stepick 1993; Grenier and Stepick 1992; see also Chapter 3 of this volume). As a result, only 28 per cent of the Census tracts in Dade have fewer than one-fifth Hispanics and the Hispanic dissimilarity index in 1990 was only 50 (Boswell 1993). Consequently, unlike Blacks, we would expect Hispanics to encounter no greater barriers to post-hurricane relocation than Anglos.

ANALYSIS OF RELOCATION PATTERNS

To determine whether race has had a significant effect on post-hurricane relocation patterns, we need to consider or control for a variety of competing explanations. For this purpose, we use logistic regression, predicting whether a household relocated. While logistic regression predicts the log-odds of house-hold relocation, the resulting coefficients can be converted, by finding the antilog, so that they tell us how much each variable increases or decreases the odds of relocation. For simplicity of exposition in the tables which follow, we present only these converted antilog coefficients along with a note of the level of statistical significance. Of course, only those variables that are statistically significant can be said, with a high degree of confidence, to have an effect on relocation.

196

In the discussion that follows, we draw on data from the three surveys we conducted at various intervals after Hurricane Andrew. Although this complicates our presentation, we believe that the replication of tests across three samples differing in size, measures, timing, and location will contribute to confidence in findings that are somewhat controversial.

In particular, our focus is on whether ethnicity and residential segregation had an effect on relocation after Hurricane Andrew. To test the effect of ethnicity, we return to the now-familiar classification of households as Anglo, Black, or Hispanic. Anglos are the reference category in all regressions. To test the effect of segregation in neighborhoods that are Hispanic or Black, we classify each household according to whether it is located in an area with a high percentage of that particular ethnic/racial group. For two surveys, segregation was less optimally measured by two variables: location in an area with 25 per cent or more Blacks, and location in an area with 25 per cent or more Hispanics. Unfortunately, households could only be placed in rather large areas – postal ZIP Code areas in the Hurricane Andrew Survey data and multi-block zones in the Homestead survey data – which did not permit the identification of more compact ethnic pockets of Blacks or Hispanics. A more refined measure of segregation – location in Census blocks with 75 per cent or more Blacks or Hispanics – could be generated with the South Dade study data.

Additional factors that could affect moving or relocating after the hurricane are also included in the analysis. We included a control for mobile homes because these were completely destroyed, resulting in almost 100 per cent abandonment. Residence in a multi-family unit was controlled for because these had higher vacancy rates than single-family dwellings one year after the hurricane. In many cases, rebuilding forced residents of entire buildings and even complexes to relocate even when their individual units were habitable. We included home ownership, although its association with post-disaster mobility is subject to differing interpretations (Morrow-Jones and Morrow-Jones 1991; Belcher and Bates 1983: 125).

Homeowners' insurance is included in the analysis because this would be expected to facilitate relocation by directly meeting additional living expenses, as some homeowners' policies did, or indirectly by insuring greater long-run household economic certainty. It must be recalled however, that the literature has consistently found that Black households are less likely to have coverage to meet additional or temporary living expenses (Bolin and Bolton 1986). Also, Black households in general, and those residing in Black neighborhoods in particular, were less likely to have been insured or to have received sufficient settlements, as we showed in Chapter 9. Since homeowners' insurance, but not renters' insurance, may cover additional living expenses, we distinguish between them in the analysis.

Income would also be expected to increase the opportunity for post-hurricane relocation and therefore is included in these analyses. In the analysis of the South Dade data however, household income was not available and a proxy variable for income – average rents for the block according to the 1990 Census – was substituted.[4] For both income and rent variables, the bottom quintile is entered as a dummy variable because empirical experimentation bore out our suspicion that the relationship was not linear. Specifically, compared to most households, those on the economic margin, in the lowest income quintile, were the least likely to move.

Two measures of damage are included. First, the respondents were asked to assess the extent of hurricane damage to their homes. For two of our surveys, this information is coded into three dummy variables indicating moderate damage, heavy damage, or complete destruction. An exception is the analysis of data from the Hurricane Andrew Survey in which moderate or more severe damage is coded into a single dummy variable. Little or no damage is the reference category for data from all three surveys. Second, overall neighborhood damage is measured indirectly by locating each respondent according to the zones used in Table 10.1 (defined by Figure 9.1 and slightly modified by splitting Zone 2). Zone 4, which is just outside the hurricane eye wall serves as the reference category. Zone 1, which is demarcated by the path of the hurricane's eye, was the most devastated.

Finally, it should be noted that the dependent variable in our analysis – household relocation – has a somewhat different definition for each survey. When analyzing the Hurricane Andrew sample, we examine whether the respondent's *household relocated outside of the hurricane impact area*. We focus on whether the *entire household temporarily left its current South Dade residence* in the South Dade Study analysis, and in the Homestead survey our focus is whether the respondent's *household, now living in Homestead, left a South Dade residence for three months or more*. In these analyses, because we are interested in exploring relocation as a response to the hurricane, only households who gave hurricane-related reasons for moving were counted as having relocated.

The results of the logistic regression analysis for each of the three surveys are presented in Table 10.3. The first equations for each are basic models which exclude segregation effects. After holding damage and other factors constant, the odds of relocating after Hurricane Andrew were significantly less for Blacks than for Anglos. The odds for Blacks ranged from approximately one-third to slightly more than one-half the odds for Anglos. In none of the equations were the odds of relocation between Anglo and Hispanic households significantly different.

After additional control variables and segregation measures are added to the models a consistent pattern emerges for households located in Black areas. The results reveal that households in predominantly Black areas had odds of post-disaster relocation that were 27 to 58 per cent lower than from other areas, depending on which additional variables are controlled. The effect of living in segregated Black neighborhoods is consistently significant and negative, regardless of which additional control variables are added to the equations.

Do the effects of residential segregation entirely account for the lower odds of Black household relocation shown in the basic models? For two of the three

Table 10.3 Logistic regression models predicting post-hurricane relocation

Models:	Hurricane Andrew Survey — Left South Dade				South Dade Pop. Impact Study — Left home			Homestead Housing Survey — Left 91 days or more		
	One	Two	Three	Four	One	Two	Three	One	Two	Three
Racial/ethnic status:										
Black	0.36[b]	0.39[b]	0.32[b]	0.43	0.52[a]	1.13	1.19	0.36[a]	0.81	1.03
Hispanic	1.04	1.07	1.04	1.14	0.92	1.05	1.20	0.63	1.05	1.08
Areal characteristics:										
25–75% Black		0.31[b]	0.27[a]	0.36[c]		0.43[a]	0.58[b]		0.39[b]	0.36[b]
75–100% Black						0.25[a]	0.42[b]			
25–75% Hispanic		0.46[b]	0.42[b]	0.54		0.87	0.93		0.29[b]	0.29[b]
75–100% Hispanic						0.76	0.79			
Housing characteristics:										
mobile home	5.91[a]	3.85[b]	4.24[b]	10.72[a]	1.91	1.86	2.93[b]	0.57	0.29	48.49
multiple units	3.40[b]	3.75[a]	3.88[a]	2.76[b]	0.77	0.89	1.35	1.51	1.40	1.51
Household characteristics:										
homeowner	0.68	0.70	0.35	0.26				0.84	0.57[b]	0.56
low income/rent				1.01						0.60
insured owner			1.10	1.78			0.41[a]			
insured renter			0.11[a]	0.20[b]			2.17[a]			
Damage/zone location:										
moderate damage					1.60	1.70[b]	1.52	8.69[a]	8.92[a]	6.71[a]
major damage	6.04[a]	6.87[a]	7.61[a]	6.74[a]	4.91[a]	5.27[a]	5.11[a]	34.30[a]	36.52[a]	27.85[a]
totally destroyed					33.82[a]	37.22[a]	32.66[a]	142.00[a]	183.54[a]	142.65[a]
eye wall	3.69[a]	6.27[a]	6.06[a]	1.57[b]	1.57[b]	1.76[a]	2.14[a]			
eye	3.58[a]	3.59[a]	3.65[a]	4.41[a]	1.48	1.73[b]	2.39[a]			
N	486	486	485	427	1,189	1,189	1,123	635	612	546
-2LL	371.1	365.4	353.2	305.4	1,332.1	1,308.7	1,212.4	623.4	556.1	542.6
Chi-square	94.7	100.3	111.2	97.4	195.3	218.7	232.0	172.4	208.4	186.8

Notes: a p ≤.01
b p ≤.05
c p ≤.10

samples, the South Dade and Homestead surveys, the addition of the segregation controls eliminates any statistically significant relocation differentials between Blacks and Anglos. And yet, in the Hurricane Andrew Survey results, Blacks continue to show lower odds of flight than Anglos in all but the final equation, despite the inclusion of segregation measures. It is possible that this significant differential for the Hurricane Andrew Survey is an artifact of the race variable picking up some of the effect of segregation, which was inadequately captured by a measure based on postal ZIP Code areas rather than the smaller, more homogeneous Census blocks. On the other hand, only the Hurricane Andrew Survey included households that left South Dade and never returned, a preponderance of which were Anglo. Consequently, the Hurricane Andrew Survey may well provide a better assessment of Anglo relocation, and therefore better capture differentials between Anglo and Black households even when controlling for segregation.

Parallel to the effect of Black segregation, living in areas with 25 per cent or more Hispanics has a significant dampening effect on relocation in two of the Hurricane Andrew and Homestead studies. However, these effects are not significant in the South Dade study in which Census blocks are used, nor in the final equation for the Hurricane Andrew Survey. Since the results across all samples suggest that Hispanic households were no less likely than Anglo households to leave, it is not clear why living in Hispanic areas would reduce post-hurricane relocation.

Clearly, more is happening than is being accounted for by our analysis. One factor that may need to be considered is the heterogeneity among Hispanics in Dade. It is possible that Mexicans experience segregation effects similar to Blacks for example. Furthermore, Mexicans are more likely to be recent arrivals, and are thus living in more segregated areas. The stronger effects of Hispanic segregation in the Homestead study support this explanation, since Hispanics there are more likely to be from Mexico.

In addition to the effects of ethnicity and segregation, the results for the South Dade survey, the largest of the three samples, show the expected tendency for low income to diminish the odds of relocation, and for having homeowners' insurance to increase the odds. Also, homeownership may have a weak inhibiting effect on post-hurricane relocation. As expected, the results for the Hurricane Andrew Survey indicate that residents in mobile homes and multi-family units were more likely to relocate than those living in single-family residences. Needless to say, higher levels of damage significantly increase the odds of relocation in virtually all analyses.

Effects on Black neighborhoods and households

Our results suggest that post-hurricane relocation was impeded for Blacks, whereas for most Hispanics, we tentatively conclude, this was not a major problem. The limited options for Blacks seeking to flee the most damaged areas of South Dade are captured to some extent in our Hurricane Andrew Survey data. While the numbers are quite small, the few Blacks (n=4) who reported relocating to North Dade moved exclusively to ZIP Code areas that were more than half Black (see Table 10.4). In contrast, virtually all Anglo and Hispanic respondents moving north of the impact zone filled vacancies in neighborhoods that were overwhelmingly White (ZIP Code areas with less than 16 per cent Blacks) – which constitute the lion's share of areas. In sum, our findings suggest that Blacks were less likely than Anglos to relocate after the hurricane not only because of possible economic constraints, but because of barriers created by residential segregation.

Table 10.4 Ethnic differences in where South Dade residents relocated[a]

	Anglo % N=33	Black % N=6	Hispanic % N=23
Where relocated			
Between Kendall Drive and Flagler Street (Zone 4)	27.4	0.0	43.7
In Dade County above Flagler Street (Zone 5)	29.8	87.8	47.3
Outside Dade County	42.8	12.2	9.0

Source: FIU Hurricane Andrew Survey
Notes: a Based on zipcode at the time of Hurricane Andrew and Zip Code at the time of the survey (12/92)
    b Chi-square = 13.4, p≤.01

BROADER IMPLICATIONS OF THE RESULTS

The findings of this study reinforce the view that residential segregation may be considered as a cause, and not simply a consequence, of racial disparities in opportunity. Several proponents of this view argue that segregation multiplies the effects of economic deprivation – poverty, crime, family dissolution, welfare dependency, and so on – by focusing them in Black, underclass neighborhoods (Wilson 1987; Logan and Molotch 1987; Squires 1991; Massey and Denton 1993). These "concentration effects" may lead neighborhoods to deteriorate beyond thresholds from which they can rebound after an economic shock (such as a recession), insuring a downward spiral of decay, disinvestment, and neighborhood abandonment.

Parallel to segregation's deleterious economic effects, which inhibit Black efforts to escape from poverty and all of its consequences, segregation would appear to limit movement out of disaster areas and in other ways stifle disaster recovery for Blacks. This is evident from a tour of Black neighborhoods in South Dade, almost five years after, which will reveal homes that have never been repaired and vacant lots on which nothing new has been built because insurance was lacking or inadequate (Chapter 9). However, less visible than the hurricane's physical destruction are social and personal losses to households in segregated neighborhoods. These households, particularly those already experiencing a precarious standard of living, may have been more susceptible to being pushed over the edge when subjected to the hardships and psychological stresses of not being able to escape hurricane-ravaged neighborhoods. Remaining trapped in a vast devastated area with no functioning businesses, services, power, or water is bound to have increased household tensions, the prospect of marital disruption, and the difficulty of finding, getting to, and keeping a job (Chapters 7, 8, and 11). For the Black community, these losses – similar to many damaged or destroyed buildings – may never be recovered.

Long-term demographic effects

Beyond the issue of ethnic differences in post-hurricane relocation and recovery, there is the question of the long-run impact of Hurricane Andrew on South Dade's ethnic composition. A demographer colleague, Lisandro Pérez (1992) argues that Hurricane Andrew will accelerate Anglo relocation, increasing the pace of *Latinization*. This argument is consistent with the notion that disasters often accelerate pre-impact trends (Prince 1920; Bates *et al.* 1963). Some support for this hypothesis is given by the data in Table 10.4, which show that among persons fleeing South Dade after the hurricane, the proportion that left Dade County entirely was highest among Anglos (43 per cent) compared to Blacks (12 per cent) and Hispanics (9 per cent). However, we do not know what proportion of those leaving the county will return. Nor can we anticipate the future influx.

The dynamism of South Dade makes predictions difficult. Reversing demographic projections from just one year earlier, the Metro Dade Planning Department (1994) now forecasts that growth in Dade County, and in South Dade in particular, will outstrip projections made before the hurricane. Rapid population growth is indicated by the fact that by 1994 South Dade had regained virtually all of its pre-hurricane population (Metro Dade Planning Department 1994). Future accelerated growth is anticipated, in large part due to increased foreign immigration into Dade County, which has outweighed domestic sources for the last couple of decades (Metro Dade Planning Department 1994). In addition, new up-scale development in South Dade may well attract Anglos. For example, Homestead survey results suggest that the Anglo population had rebounded by 1995 (Peacock and Gladwin 1995).

Implications of segregation trends

Regardless of the impact of Hurricane Andrew, the decline in segregation that began in the 1960s is likely to continue, leading to a more even geographic distribution of ethnic groups. The speed of this process will depend, in part, on the extent to which applicable federal law is enforced; the 1988 Fair Housing Amendments dramatically increased both the fines and prospects for legal challenges to discrimination in housing (Massey and Denton 1993). Perhaps these developments will send a signal to landlords and property sellers – as well as prospective buyers and tenants – that discrimination will bring costs that are better avoided.

To the degree that residential segregation remains a key feature of American cities, racial inequality must be considered when assessing the urban landscape's vulnerability to natural disasters. Our findings suggest that disaster research, inherently ecological in orientation, needs to focus more on urban ethnic ecology and residential segregation. Segregated neighborhoods are a part of the urban landscape that are not well integrated with the mainstream institutions critical for jobs, political power, financing, and insurance. For this reason, Blacks are likely to be more vulnerable to disasters and to experience more barriers to recovery. Natural disasters may simply reinforce the effects of segregation and marginalization, isolating ghettoized Black communities further from the market mechanisms that underlay recovery and upward mobility in most industrial societies.

Ultimately, then, when speaking of a "natural" disaster, we must not forget that when the wind howls, water rises, or the earth trembles, the real disaster unfolds only when *social systems fail* because of inadequate preparation for these natural occurrences. That is, a disastrous outcome can be averted by having safe shelters, adequate building codes, evacuation plans, and so on. Our examination of the South Dade recovery process suggests that residential segregation contributes to the potential for the failure of social systems by maintaining barriers to post-disaster relocation, thereby compounding the problems of vulnerable communities. Unfortunately, given the slow pace of residential desegregation over the past 30 years, these high-risk areas are not likely to disappear as a result of market forces alone. Moreover, the current political climate in the United States would not lead us to expect swift state intervention to correct these problems. Therefore, it behooves disaster planners and responders to anticipate the vulnerability of highly segregated communities and to develop special mitigation and response initiatives to address their special circumstances.

POLICY IMPLICATIONS

The above discussion suggests a number of policy implications:

• A policy implication directly implied by the above is that issues related to the maintenance and creation of segregation must be further examined and addressed long before a disaster strikes, because they represent potential weaknesses in the ability of our communities to recover. Proactive solutions must not only include the active pursuit of legal remedies to combat segregation such as fair housing laws, but also the rigorous utilization, where necessary, of testing programs to gauge the on-going recreation of segregation by real-estate and mortgage companies via racial steering, red-lining, and other mechanisms. Without such actions these more hidden and insidious factors contributing to racial segregation will continue or reoccur, particularly in communities with longstanding patterns of racial and ethnic segregation.

- While risk assessment and mapping is an important part of any community preparation and planning, vulnerability mapping, which takes physical risks a stage further by overlaying them with social factors that produce vulnerability, is a necessary step in building effective community mitigation. Patterns of existing segregation, for example, must be clearly mapped out as part of a community's vulnerability assessment. If areas of high minority concentrations, defined in economic, social or cultural terms are identified, additional actions such as detailed reconnaissance studies should be implemented to identify potential problems, develop solutions, and invest resources that will mitigate against disaster impact and subsequent problems. Policy initiatives stemming from detailed community vulnerability studies must not have as their simple goal the problems of relocation, but rather the overall goal of mitigation as part of normal community development.
- While not directly related to this analysis, the results suggest that Federal Emergency Management Agency programs such as temporary housing and rent vouchers should be fully examined and evaluated in terms of access, utilization, and effectiveness. The variable rates of relocation seen following Hurricane Andrew, especially in the face of the high levels of destruction, suggest that Federal programs to assist in relocation may have been less than effective, particularly for Black households. It may well be that Black households did not have full access to these programs due to self-selection, or a failure of these programs to reach into predominantly Black areas. Also, it is possible that in the face of discrimination Black households may have experienced constrained housing options, even with rent vouchers in hand. These are all issues that must be further explored, and where necessary legal and policy solutions developed.

NOTES

1   Note that in this analysis we have made one change in the zones identified in Figure 9.1; we have divided Zone 2 into two parts, divided by SW 136 Street.
2   Several alternative possibilities could account for the changes in South Dade's ethnic mix. Between 1990 and 1993 there may have been non-storm related growth in South Dade's Black and Hispanic populations and/or decline in Anglos. The latter would follow a trend in *White flight* over the last several decades (Boswell and Curtis 1991). Unfortunately, there are no data on ethnic composition of South Dade for 1991, the year before the storm, nor 1992, the year of Andrew, to use to evaluate this possibility. However, the time period under consideration is relatively narrow and Dade did not experience a major influx of immigrants during it. Thus, while "normal" White flight may explain some of these losses, it is unclear if this alone would account for such a large decrease in the Anglo population.

Yet, another possibility is that the differences are simply an artifact of the data collection, which resulted in Blacks being slightly over-represented in our sample. A test of this hypothesis was conducted by generating a 1990 Census estimate from the South Dade Population Impact Study data by matching respondents with the per cent Black in their block. This percentage was then weighted by the inverse of the probability of selecting that respondent and a new estimate was derived by summing across all respondents. The weighting was necessary because the survey was conducted using a disproportionate stratified sampling technique with the probability of any block being selected equally to its proportionate size. So each Census block had a different weight. Recalculating a 1990 estimate in this fashion generated an estimate for the proportion Black that was 2 percentage points higher than the 1990 Census figures. This higher estimate, coupled with a slightly elevated (3 per cent) response rate for Census blocks with more than 75 per cent Blacks, may account for the higher per cent Black in the results presented in Table 10.1. However, the data do not uniformly support this interpretation. For example, response rates and the estimated 1990 Black population were not higher in Zone 2B. And yet, Table 10.1 shows that Zone 2B registered a large increase in Blacks between 1990 and 1993. Furthermore, deriving a 1990 estimate for the percentage Hispanics in the 244 sample blocks was consistent with 1990 Census figures. On the whole then, Black gains do not appear to be simply an artifact of the data or data collection procedures.

3   $D = \Sigma \mid P_1 - P_2 \mid /2 \times 100$, where P_1 is the proportion of the first ethnic group's total population (for example in Dade County or South Dade) on each block, and P2 is the analogous proportion for the second ethnic group (note that $\Sigma P_1 = \Sigma P_2 = 1$). For calculations in this chapter, non-Hispanic Whites and non-Hispanic Blacks were used at the tract level, whereas Whites and Blacks (both categories include Hispanics) were used at the block level.

4   For blocks for which rents were not given, average rents were predicted using median home values and other block characteristics. Predictions were based on a regression analysis for all blocks in South Dade, each block being weighted by the number of housing units it contained.

11

AND THE POOR GET POORER

A neglected Black community

Nicole Dash, Walter Gillis Peacock,
and Betty Hearn Morrow

In their award-winning book, Portes and Stepick (1993) characterized metropolitan Miami as a *City on the Edge*. The vision of a city on the edge simultaneously captures the dynamics and excitement, as well as the danger and uncertainty, facing an urban area that is the essence of a real-world multi-lingual, multiracial, and multicultural social experiment. The focus of this chapter is a very small, predominantly Black incorporated community – Florida City – which is embedded, albeit peripherally, within the complex urban ecological network that is metropolitan Miami.

Nowhere were the effects of Hurricane Andrew more strongly felt than in Florida City. As a community, it provides a case study of how incorporated pockets of neglect at the lowest rungs of an urban stratification system can have as much to do with disaster difficulties as the event itself. As we shall document, in comparison to its immediate neighbor of Homestead, Florida City was unequally impacted by the hurricane; its recovery progress has been more sluggish; and its future appears more tenuous. Prior to Hurricane Andrew, Florida City was also a city on the edge – in this case the edge of poverty – and Hurricane Andrew may well have blown it over the brink.

URBAN ECOLOGICAL NETWORKS, MINORITY COMMUNITIES, AND DISASTER

The dynamics of urban ecological networks are a function of the distribution of power and influence, as well as the distribution and flow of resources (Chapter 2). Political and business elites typically wield disproportionate power over economic development and public policies related to services, infrastructure, zoning and land use, and taxes. It is not a coincidence that decisions over access to scarce resources disproportionately reside with these same individuals and groups. This does not mean that a single omnipotent elite group controls the network, as suggested by power elite models (cf. Hunter 1953; Mills 1956; Dahl

1961). Rather, centers of control and influence shift, emerge, merge, and decline. They are multifaceted – based on coalitions, conspiracies, and cooperative ventures negotiated and renegotiated between individuals, groups, and organizations that at other times may be in conflict and competition (Nix, Dressel and Bates 1977). Today's coalitions hold only until issues emerge in which members' interests do not coincide. Nevertheless, the concerns of banking, real estate, and builders, for example, tend to be in concert where *development and growth* issues are concerned. An analysis of this process is critical for understanding the urban area as a "growth machine" (Logan and Molotch 1987).

Local government is often touted as the mechanism for enhancing a community's sense of identity and providing control over its development. This belief was clearly evident in the incorporation fever that spread throughout metropolitan Dade County following Hurricane Andrew (Hartman 1994; Dluhy 1995; Williams 1995). Interestingly, this fever occurred not just in affluent Anglo neighborhoods, but also in some predominantly Black areas (Walters 1995a; Walters 1995b; Walters and Swarns 1995). It is likely, however, that incorporation will prove to be not in the best interests of the latter. Small, economically and politically weak governmental entities can be easily manipulated by powerful business elites and can actually create and perpetuate inequality (Hill 1974; Newton 1976).

The presence of small local governments within a stratified urban ecological network can be both a result and a cause of continuing class and racial segregation. At best, working class and minority towns are often forgotten or ignored communities containing inhabitants and property that few would claim; at worst, they become the foundation for a "cycle of poverty," rationalized or justified by the deficiencies of their inhabitants (Logan and Molotch 1987: 197). With few internal resources, leadership in these communities is highly dependent on and susceptible to external influences. Unfortunately, the effects of the larger socio-political context are rarely discussed when minority communities fail in the development game. Instead, blame is typically attributed to the "types of people" living there or the failings of ineffectual community leaders.

In a disaster context, a well-organized and economically strong local government with good leadership, knowledge, and the power to act has been found to promote successful community recovery (Klinteberg 1979; Rubin 1985). The political and economic realities of today's new federalism carry the expectation of increased local responsibility that heightens the need for effective and strong leadership. However, given the structural difficulties small minority and low-income communities can experience during normal periods, would they expect to fare any better during post-disaster situations?

Minority and poor households tend to be more adversely impacted by disasters (Chapter 9; Cochrane 1975; Perry and Mushkatel 1986; Streeter 1991) and encounter more recovery problems (Moore 1958; Bolin 1986; Bolin and Bolton 1986; Bolin and Stanford 1991; Phillips 1993b). What is less clear is how larger potentially discriminatory structures and processes impact a minority or impoverished local government's ability to garner resources to respond. Peacock, Killian, and Bates (1987) found that households residing in peripheral isolated rural villages recovered more slowly than households residing in more complex, economically and politically central cities (see also Peacock *et al.* 1984; Hoover and Bates 1985). Households within larger communities had better access to external resources and were able to take advantage of post-disaster funding and resource opportunities (Bates and Peacock 1987). The implication is that a community's position within a regional stratification system and exchange network has important consequences for disaster-related processes. The case study of the impact of Hurricane Andrew on Florida City draws attention to the relevance of the larger socio-political context. By applying a socio-political ecology perspective and comparing Florida City to its more affluent neighbor of Homestead, we highlight ways in which an *entire community* can be at risk of failing to recover.

FLORIDA CITY

Florida City is the third oldest municipality in Dade County and is also one of the smallest – a pre-storm population of approximately 6,000 persons in an area of only 2.5 square miles. Located 35 miles south of the City of Miami in southwestern Dade County, it is bordered by the City of Homestead to the north, Biscayne Bay to the east, and the Everglades National Park to the west. The gateway to the Florida Keys, it is the last community before the Overseas Highway that ends in Key West (see Figure 1.2, page 7). Like so many South Florida communities, its origins are tied both to agriculture and to the historic figure Henry Flagler (Taylor n.d.). In the early 1900s Flagler enticed a central Florida farmer to experiment with growing tomatoes in the area south of Homestead. The effort met with limited success – what is today considered prime agricultural land was then covered with water most of the year. A canal was eventually dug to drain the land, making it viable for farming. By 1912 over 300 families lived in what was then called Detroit, and its business area was larger than that of its neighbor, Homestead. When residents voted for incorporation two years later they renamed the town Florida City. There was an attempt to consolidate with Homestead in 1936, but Florida City residents, mostly White farmers, voted it down.

Many changes have occurred in Florida City since then. According to the 1990 Census, it is now 61 per cent Black. Another 37 per cent of its residents are Hispanic, mostly Mexicans and Central Americans drawn to the area for agricultural work. Of the 1,782 housing units reported, 55.3 per cent were rentals, mostly single-family homes or duplexes. Only 17 per cent of Florida City's housing stock was classified as multi-unit structures. While remapping areas of Florida City as part of the South Dade Population Impact Study, our field team was careful not to go by outward appearances. On a number of occasions "single-family" structures had been subdivided, with each room being rented to individuals, families, or households sharing common kitchen and bathroom facilities. The condition of these residences often reflected chronic ill-repair. The average 1992 real property value in Florida City was $27,628 for a single-family home, $28,536 for a duplex unit, and $68,270 for a multi-family building. These property values are extraordinarily low for the Miami area; however, economic factors nevertheless preclude most residents from home ownership. The median household income in Florida City was only $15,907 in 1990.

While its demographics have changed, Florida City's economy has remained highly dependent on agriculture. Agricultural sales accounted for nearly 60 per cent of the city's 1992 sales volume, over four times more than retail sales. The majority of its 2,678 employed residents were in agriculture or transportation/ communication, many of the latter undoubtedly involved in transporting agricultural products. These Census figures actually underestimate the centrality of agriculture because they exclude migratory labor. Florida City is economically and politically dominated by large agricultural interests – to the extent that agri-businesses pay any attention to local concerns.

Florida City's economy is very small and lacks diversity, which may account for its 16.4 per cent unemployment rate. It is one of the five poorest munici-palities in the United States (City of Florida City 1993). As a poor community of minority residents, the town contains only 0.3 per cent of Dade's population. It is no wonder that Florida City's government wields negligible power in the larger political context. With a relatively small residential tax base and a very limited economy, the municipality is highly dependent upon both state and federal resources. Furthermore, these funds are usually filtered through Metro Dade County. Florida City is the epitome of a peripheral, highly dependent, incorporated community located within one of the nation's largest metropolitan areas.

HOMESTEAD

Throughout much of this discussion Florida City will be compared to its neighbor of Homestead. The two communities are geographically contiguous and were the only two incorporated political entities in South Dade at the time of Hurricane Andrew. Both were directly in the storm's path. While similar in political designation and location, they are quite different in size, racial compo-sition, median income, and housing value. In 1990, Homestead had a population of 26,866, nearly five times that of Florida City. Homestead's 9,317 households averaged 2.83 persons, compared to Florida City's 3.17. The median income of Homestead's residents was $20,594 and the average home value was $32,313, both approximately 20 per cent higher than Florida City's.

Homestead was one of the most Anglo communities in Dade County – over 42 per cent of its population was classified as non-Hispanic, non-Black. Hispanics made up 35 per cent of the population and included many Cubans as well as Mexicans. Homestead's predisaster economy was larger and more diversified; total sales were roughly seven times higher. The top business activities were retail trade, professional services, construction, and FIRE industries (Finance, Insurance, and Real-Estate). At 7.3 per cent, the unemployment rate was less than half that of Florida City. Homestead's economy and influence were also enhanced by the presence of a US Air Force base within its city limits.

These were the two communities located directly in the path that Andrew carved across the Florida peninsula on 24 August 1992. While both were small, rather rural, incorporated towns, they faced the storm and subsequent recovery from different positions, both in terms of their internal characteristics and access to external resources. In many respects, Homestead became the symbol of post-impact destruction and suffering for the entire area. Its city manager, Alex Muxo, and Homestead Air Force Base were featured prominently on the national television coverage. While Florida City also appeared in aerial surveys and many photos, it was rarely mentioned by name.

UNEQUAL CONSEQUENCES

Our discussion of Hurricane Andrew's impact will focus on three interrelated spheres: housing, population, and the economy. Following conventional community studies (cf. Friesema *et al.* 1979; Wright *et al.* 1979), aggregate level statistics are employed to compare the recovery progress of Florida City with that of Homestead. Our sources include Metro Dade County tax rolls, Dun and Bradstreet business information, and primary data from our South Dade Population Impact Study. Impact assessments concentrate on relative and absolute changes in housing, economic, and population characteristics between 1992 and 1993, allowing for an assessment of loss.[1]

Impact on housing

According to city officials, every building in Florida City was damaged, an assessment supported by the data presented in Table 11.1. In all, Florida City's aggregate residential building value decreased from $38,966,559 in 1992 to $8,626,626 after the storm – a dramatic 78 per cent reduction in residential property value. The greatest decrease occurred among single-family homes, which lost 79 per cent of their average value. While Homestead's total residential loss was almost twice that of Florida City, its relative loss was smaller – about 60 per cent of its aggregate value.

Another way to compare the relative losses between the two communities is to calculate a *value comparison ratio* of property values before and after impact. In 1992, a single-family home in Homestead had an average value of $32,313 while in Florida City it was $27,628, yielding a ratio of 1.17 to 1. In other words, for every $1.17 of single-family housing value in Homestead, Florida

Table 11.1 Comparison of residential property loss in Florida City and Homestead after Hurricane Andrew

Florida City	Single family	Duplex	Multi-family	Total
Average value 1992 ($)	27,628	28,536	68,270	38,966,559
Average value 1993 ($)	5,770	6,789	18,416	8,626,626
Average value lost ($)	21,858	21,747	49,854	30,339,933
Percentage change	−79	−76	−73	−78
Homestead	*Single family*	*Duplex*	*Multi-family*	*Total*
Average value 1992 ($)	32,313	34,453	144,214	134,620,731
Average value 1993 ($)	12,721	17,096	49,756	52,718,467
Average value lost ($)	19,592	17,357	94,458	81,902,264
Percentage change	−61	−50	−65	−60

Value comparison ratio (Homestead/Florida City)				
	Single family	*Duplex*	*Multi-family*	*Total*
Pre-impact 1992	1.17	1.21	2.11	3.45
Post-impact 1993	2.20	2.52	2.70	6.11

Source: Metropolitan Dade County 1992 and 1993 real property tax data, Dade County Tax Collector

City had $1.00. If the impact of the storm was equal, this ratio should remain relatively unchanged; however, it increased across all types of housing. Furthermore, the ratio for total residential property value increased from 3.45:1 to 6.11:1 after Andrew. Similarly, while the rate of vacant residential units was high for each city in the summer of 1993, about one year after the storm, it was over 54 per cent for Florida City, compared to about 40 per cent for Homestead. In light of the extreme damage sustained by each community, it is awkward to make distinctions as to which town suffered most. Nevertheless, the effect on the housing stock of Florida City was more severe, as measured by residential property values, and the gap between the two communities widened.

These numbers do not adequately capture the impact of the storm on individual families and households. During fieldwork for the South Dade Population Impact Study we observed many people living in severely damaged homes for lack of viable alternatives. While officially condemned and without services, we discovered that many badly damaged dwellings continued to be occupied. Families lived for months in houses with damaged roofs, no electricity, and often without running water. Adding to the discomfort and health hazards, each day brought temperatures in excess of 90 degrees Fahrenheit and heavy seasonal rain to a landscape stripped of its normal sub-tropical covering of shade trees. As one respondent noted, "I wouldn't be living here if I had anywhere else to go."

Population changes

As would be expected, the massive destruction resulted in many people leaving the area. While 1994 estimates suggest that the South Dade area experienced a loss of approximately 67,000 individuals, or about 18 per cent of its 1990 population base (Smith and McCarty 1995; Smith 1994), both Florida City and Homestead had much higher relative losses. Florida City lost an estimated 33 per cent of its population, or about 1,978 people. Homestead's population loss was slightly lower at 31 per cent, or 8,355 individuals. In both cases, more Anglos relocated out of the area than did Blacks. The Black proportion of Florida City's population increased by more than 10 percentage points to 71.4 per cent, while its Anglo population fell approximately 5 points to 16.5 per cent. In Homestead, Blacks gained 6.5 points to 29.6 per cent, while Anglos lost 8.6 points to 33.2 per cent. Interestingly, the Hispanic population actually fell in Florida City to slightly more than 12 per cent and increased slightly in Homestead to 37.3 per cent of the total. The net effect was that Florida City became an almost exclusively Black community while Homestead was left with relatively equal proportions of each group, with Anglos losing their numerical majority.

Economic effects

The economy of any community is critical to its vitality and the well-being of its residents. In the context of a small community contained within a large metropolitan area, the strength of its local economy is crucial to its ability to influence regional policies. Using 1992 and 1993 tax assessments, comparison of the value of commercial properties in the two towns reveals little overall difference. As seen in Table 11.2, both Florida City and Homestead suffered substantial losses, particularly among properties used for sales, repairs, entertainment, and lodging. As a result, Florida City's average commercial property value dropped 31.5 per cent, compared to a decrease of 29 per cent for Homestead. However, a comparison of commercial property value alone does not adequately reflect the economic impact of Hurricane Andrew on Florida City.

If we compare the changes in number of businesses, number of employees, and sales revenues, a more complete picture emerges. According to the Dun and Bradstreet data reported in Table 11.3, Florida City had 114 businesses in 1992. A year later the number had fallen to 55, with losses across all segments except construction and manufacturing. The two hardest hit areas were Florida City's two largest employers: agriculture and the retail trade. In comparison, Homestead had 797 businesses in 1992 and 777 in 1993. However, some gains were registered in a number of areas, including business services and agriculture. A year after Andrew, Homestead had experienced a net loss of 2.5 per cent while Florida City had lost 52 per cent of its pre-storm businesses.

Table 11.2 Comparison of commercial property loss of Florida City and Homestead after Hurricane Andrew

| Property use | Florida City | | | Homestead | | |
	1992 value ($)	1993 value ($)	% change	1992 value ($)	1993 value ($)	% change
Retail/wholesale						
sales	3,892,377	2,137,483	−45.1	33,113,583	19,454,568	−41.2
Repairs	165,332	51,346	−68.9	688,358	485,330	−29.5
Office building	724,654	119,219	−83.5	10,637,984	9,653,248	−09.3
Entertainment	132,420	5,683	−95.7	1,801,282	118,133	−93.4
Auto/marine	941,650	295,395	−68.6	3,426,049	2,247,384	−34.4
Hotel/motel	5,916,344	3,568,816	−39.7	5,072,696	2,620,134	−48.3
Service station	814,226	690,292	−15.2	860,158	667,382	−22.4
Manufacturing/						
industry	6,051,680	4,710,636	−22.2	8,830,306	4,744,699	−45.4
Government	2,897,079	2,657,094	−08.3	11,535,211	11,108,512	−03.7
Club/education	277,934	274,738	−01.1	657,989	640,655	−02.6
Cultural/						
religious	1,358,032	1,304,953	−03.9	6,161,686	4,517,879	−26.7
Utility	1,135,722	1,108,791	−02.4	788,064	4,058,479	415.0
Miscellaneous	1,073,605	452,198	−57.0	12,156,440	7,657,702	−37.7
Total	25,381,055	17,376,644	−31.5	95,729,806	67,974,105	−29.0

Source: Metropolitan Dade County 1992 and 1993 real property tax data, Dade County Tax Collector

In terms of jobs, the consequences were even more dramatic. Again using Dun and Bradstreet data, Table 11.4 compares the number of employees in 1992 and 1993. Overall, Florida City lost approximately 87 per cent of its employees. The most dramatic losses were in transportation/communication (−99.7 per cent), agriculture (−92.4 per cent), and retail trade (−84.4 per cent). In contrast, Homestead gained 10 per cent to its employee base, mostly in agriculture, retail trade, and professional services. Homestead did, however, register losses in construction and manufacturing and Homestead Air Force Base, which is estimated to have cost the area 8,000 jobs (Office of County Manager 1994). The Department of Defense has decided to rebuild, but with a vastly scaled-down presence. Currently, the county is still planning for the development of the remainder of the base, probably to divert some air-cargo traffic away from Miami International Airport.

In light of the loss in businesses and employees, it should not be surprising that Florida City had an astounding 83 per cent drop in sales between 1992 and 1993. The data presented in Table 11.5 show losses in almost every area. Sales in agriculture, its leading sector, fell 93 per cent. Wholesale and retail trade each experienced an 84 per cent reduction in volume. Homestead's total sales volume, on the other hand, fell by only 1.1 per cent, but not because there were no

Table 11.3 Comparison of number of businesses in Florida City and Homestead before and after Hurricane Andrew

Type of business	Florida City			Homestead		
	1992	*1993*	*Change*	*1992*	*1993*	*Change*
Agriculture	7	2	−5	57	59	+2
Construction	6	6	0	110	88	−22
Manufacturing	5	5	0	26	23	−3
Transportation/ communication	2	1	−1	23	25	+2
Wholesale trade	5	2	−3	52	50	−2
Retail trade	28	10	−18	187	184	−3
Finance/insurance/ real-estate	5	4	−1	65	65	0
Business services	32	12	−20	116	123	+7
Professional services	20	11	−9	153	149	−4
Public administration	4	2	−2	8	11	+3
Total	114	55	−59	797	777	−20
Percentage change		−52			−3	

Source: Dun and Bradstreet Business Data from Conquest

Table 11.4 Comparison of number of employees in Florida City and Homestead before and after Hurricane Andrew

Type of business	Florida City			Homestead		
	1992	*1993*	*Change*	*1992*	*1993*	*Change*
Agriculture	432	33	−399	406	705	+299
Construction	17	19	+2	608	486	−122
Manufacturing	86	28	−58	298	242	−56
Transportation/ communication	826	2	−824	130	135	+5
Wholesale trade	30	15	−15	465	492	+27
Retail trade	270	42	−228	1,734	2,015	+281
Finance/insurance/ real-estate	34	14	−20	482	476	−6
Business services	455	28	−427	505	479	−26
Professional services	222	59	−163	1,669	1,944	+275
Public administration	306	96	−210	468	499	+31
Total	2,678	336	−2,342	6,765	7,473	+708
Percentage change		−87			+10	

Source: Dun and Bradstreet Business Data from Conquest

Table 11.5 Comparison of sales volume in Florida City and Homestead before and after Hurricane Andrew[a]

Type of business	Florida City			Homestead		
	1992	*1993*	*Change*	*1992*	*1993*	*Change*
Agriculture	43,228	3,000	−40,228	64,788	107,240	42,452
Construction	1,280	1,430	+150	84,156	34,718	−49,438
Manufacturing	6,148	2,514	−3,634	19,545	13,349	−6,196
Transportation/ communication	135	100	−35	16,263	24,611	+8,348
Wholesale trade	6,233	1,000	−5,233	53,956	84,802	+30,846
Retail trade	10,747	1,718	−9,029	165,514	158,000	−7,514
Finance/insurance/ real-estate	1,856	1,256	−600	71,643	48,648	−22,995
Business services	2,921	1,033	−1,888	18,443	15,858	−2,585
Professional services	3,637	1,136	−2,501	94,884	95,371	+487
Total	76,185	13,187	−62,998	589,192	582,597	−6,595
Percentage change		−83%			−1%	

Source: Dun and Bradstreet Business Data from Conquest
Note: a In thousands of dollars

losses. Losses were experienced in construction (−59 per cent), among FIRE businesses (−32 per cent), and in manufacturing (−32 per cent), but they were offset by gains in agriculture (+65.5 per cent), wholesale trade (+57 per cent), and transportation/communications (+51 per cent). The net effect was that Florida City's already small economy was virtually wiped out, while Homestead, with its larger and more diverse economy, fared somewhat better.

Several additional reasons may help explain why Florida City's economy suffered so disproportionately. First, any business already thinking about leaving the area may have seized this opportunity to relocate. A number of agricultural enterprises, for example, simply moved from Florida City. Also, businesses in Florida City were more likely to be marginal and were more likely to be inadequately insured and/or have insufficient resources to rebuild. Those in Homestead may simply have had relatively greater capacity to quickly regroup and rebuild.

The first businesses reopening after the storm were almost exclusively national fast-food franchises and retail chain outlets, most of which were located in Homestead. In contrast to small owner-operated businesses, they could access multinational and national capital resources to quickly get back in business. These explanations do not sufficiently explain the differences in recovery rates between the two municipalities, however. To understand better the circumstances surrounding the recovery of each city, we need to examine the efforts undertaken by their respective governments, as well as by private and semi-private agencies involved in recovery assistance.

Plate 11.1 Some national franchises were up and running even before the rubble
was cleared
Source: *Miami Herald* staff photo

THE ROAD TO RECOVERY?

The complexities of *community* – composed as it is of households, businesses,
and governmental organizations – demand a comprehensive view when seeking
to understand the dynamics of restoration and recovery. When focusing on
housing or business recovery, the role of government is central, not only to
restoring public buildings and community infrastructure, but also to marshaling
and coordinating internal and external resources (Rubin 1985; Klinteberg
1979). In this competitive and conflictual process for control over recovery and
redevelopment resources (Bates and Peacock 1992), it is especially important
that small communities – often lost in the bureaucratic morass associated with
large metropolitan areas – have a viable and proactive local government during
the scramble for post-disaster resources. Without effective leadership, responsive
to internal needs and influential outside the community, small communities'
recovery needs are unlikely to be adequately addressed.

216

Plate 11.2 A few local businesses, such as this Florida City barbershop, managed to
reopen without windows or electricity
Source: Marsha Halper/*Miami Herald*

While a storm the magnitude of Hurricane Andrew will initially leave any
community in organizational disarray, the situation in Florida City approached
chaos. The entire government complex, including the city hall and police
station, was destroyed. Even worse was the damage suffered by Florida City's
social infrastructure. The city's leadership was decimated. Several members of its
management team, including the purchasing director, city engineer, and finance
director, resigned in the immediate aftermath of the storm (Hartman 1993c).
At the same time, many departments had to expand to deal with the crisis. Over
thirty additional staff had to be hired and trained, including three new building
inspectors. Less than two years after the storm, a protracted search was conducted to replace the city manager (Arthur 1994b; Arthur 1994c). Florida City's
normal disadvantage in Dade's political and economic structures was further
crippled by its lack of experienced administrators and staff as it attempted to
deal with the complex problems of recovery.

As a Florida City official interviewed over a year after Hurricane Andrew
commented, "No one plans for recovery." This lament was undoubtedly heard
throughout South Dade, but it took on added saliency in the context of Florida
City's limited resources. In the wake of natural disasters, particularly high-
impact disasters, local communities are often presented with opportunities to
gain access to skills and resources far beyond normal levels (Bates and Peacock
1987). Communities with preexisting plans or the ability to organize effectively

are quickly able to generate grant proposals and reconstruction projects to take full advantage of the possibilities for garnering new resources. For Florida City, these opportunities were largely missed. As another official stated, "Florida City was a small city that had very little, if any, capability of administering grants. What does a disaster do to a small city that never had this capacity? It throws them into turmoil." Furthermore, this case study supports the notion that lack of planning, whatever the reason, will delay or restrict community recovery and redevelopment. The same Florida City official noted, "A lot of volunteer organizations came in, and a lot of government organizations came in, but it took more than 18 months to get organized. This makes you wonder, how do we help people to recover?"

In addition to internal problems, other difficulties and frustrations stemmed from its highly dependent position within the larger socio-political context. When plans were finally generated and proposals authorized, the painful wait began. A key administrator noted that they had received three grants:

> I have almost spent the federal money. The state grant [we] haven't gotten yet, and here it is [over] a year later. They said the check is in the mail. Well . . . I still haven't received it . . . Once you get geared up and ready to go, you are still waiting for the money.

For a community with little if any reserves, the wait for funds is extremely serious. Not only do residents suffer directly by not having necessary housing or services, they may also suffer indirectly. Vital businesses in a quandary over whether to stay or leave the area may opt to relocate when recovery is perceived to be moving at a snail's pace.

Similar frustrations were expressed concerning normal funding channels via Dade County's municipal government. Noted one official, "[the County] received money from the state and the federal government – community block grants – and the County controls the whole thing." In many ways, Florida City lacked control over its destiny. While it is an incorporated, politically autonomous municipality, Florida City relies on Metro Dade for many necessary services, such as fire rescue and tax collection. The County administers most of the public funds during normal and emergency periods, including county, state, and federal. There was a strong sentiment among Florida City residents that their community was not on the County's priority list. In fairness to Metro Dade, Hurricane Andrew crippled over one-third of Dade County. The demands for assistance were unprecedented and priorities were undoubtedly difficult to set and maintain. Those communities who were most effectively organized to use the funds were obviously in the best position to receive them.

Prior funding relationships and agreements related to the dependency of the town also added to the difficulties. According to one city administrator "the county doesn't give [Florida City] any administrative money. They keep it all for themselves." Without adequate administrative overhead, the hiring, training, and retention of city personnel is difficult. Not surprisingly, the same official suggested that "the city should be given the money and let [Florida City] . . . get the things done that need to be done. We can read rules. The county is afraid you are going to do something wrong. Well, we can handle it."

Florida City had two basic hurdles to overcome beyond the disaster itself. The first was internal: community leaders were not prepared to deal with the multitude of tasks associated with post-disaster restoration and recovery. This was exacerbated when their leadership structure was decimated. The second hurdle was external: viewed within the larger power structure of a metropolitan urban community, Florida City was a very small player. It was highly dependent upon external funding – funding that, from the perspective of Florida City, was unnecessarily delayed.

Initiatives for housing and community redevelopment

Despite problems, 18 months after Hurricane Andrew Florida City had received a total of $6.2 million from three grant programs for housing restoration and redevelopment (City of Florida City 1994). Federal funds included $3.3 million in HOME Disaster Area grants and $1.5 million in a Community Development Block Grant (CDBG). While HOME grants are specifically designed for housing, the CDBG funds can be used for a wide range of activities, including neighborhood revitalization, economic development, and improved community facilities and services. It also received a $1.4 million grant from the state Hurricane Andrew Recovery and Rebuilding Trust Fund generated from increased hurricane-related sales tax revenues. According to City officials, all of these funds are earmarked for housing projects. While 636 new housing units are proposed, the actual number will depend on how much additional funding is received.

Florida City received some unanticipated support from the private sector. In the immediate aftermath, Ed Ryan of Ryan Homes – one of the largest residential developers in Pennsylvania – decided to assist in South Dade's recovery (Arthur 1994b). He set up the Southwin Development Corporation to build affordable housing. After being rebuffed by Homestead when speedy building permits were requested, Southwin focused its attention on Florida City. The result was the Villages of Palm Bay, an $18 million, 235-unit affordable housing development. By the summer of 1994, 76 homes had been sold in Phase One and 60 were occupied (Arthur 1994a; Arthur 1994b).

Florida City was quickly recognized as an area likely to have difficulty recovering and a variety of semi-private redevelopment plans were launched. Just two months after Hurricane Andrew, a group of architects and planners held the Florida City Charrette – an intensive design and planning effort. The idea was not merely to restore Florida City, but to recreate it, allowing the community to "rise from the hurricane wreckage with a series of model neighborhoods, each with its own public park and community center, and an old-fashioned main street" (Dunlap 1992). The plan called for the development of a "pioneer village" as the centerpiece, including shops, civic buildings, a visitor's center, and a hotel.

The second initiative was a component of an overarching plan for the redevelopment of the entire South Dade area known as the New South Dade Planning Charrette (NSDPC) (New South Dade Planning Charrette 1993). We Will Rebuild (WWR) funded this charrette to stimulate alternative redevelopment initiatives. The NSDPC generated redevelopment models promoting environmental renewal, self-sufficiency, and community identity. Rather than a detailed comprehensive plan, a series of case study models and general principles were developed. As part of this process, two housing plans were developed for the Florida City area – one based on detached housing and the other on courtyard housing. Both plans attempted to incorporate culturally sensitive housing designs from Mexico and Guatemala.

City officials, including the mayor and city manager, supported the Florida City Charrette plan. The people of the community, however, were less supportive. Residents complained that the City should worry less about future economic development and more about immediate housing needs (Brennan 1992c). There were fears that redevelopment would push out long-term residents unable to afford living in the new community. Some suggested that the plan would fail to draw new, more affluent residents, but would succeed in pushing out poor residents. Many experts expressed concern that the plans were too ambitious and failed to address immediate community needs. All of these concerns impacted the eventual decision to drop the Florida City Charrette plan.

In an interesting twist, a similar redevelopment plan for Homestead also elicited negative responses from minorities almost a year later (Arthur 1995). The centerpiece of the planned Homestead project was also a pioneer village consisting of 120 middle-income homes and requiring the demolition of a farmworkers' neighborhood. Leaders from the African-American, Hispanic, and Haitian communities demanded input into the plan. During this time period, The Enterprise Foundation, a national planning and urban redevelopment firm known for its sensitivity and concern for low income housing, announced plans to assist with the planning and rebuilding of Homestead. However, when the planning team pushed for the pioneer center, The Enterprise Foundation removed its name from the project, calling it "urban removal" and decrying the lack of concern for low-income housing (Arthur 1995). Despite these setbacks, Homestead is investing $30 million of federal and state funds to implement the plan.

We Will Rebuild also provided funds to facilitate the planning and implementation of recovery and redevelopment projects in Florida City as well as Homestead (Donnelly 1993). In 1993, it allocated $236,000 to the joint Homestead/Florida City Chamber of Commerce – in part to aid microbusinesses. Florida City's government received $185,000 to pay for technical assistance and an additional $97,000 to repair and improve two city parks. WWR funds were also allocated to projects located in Homestead, including approximately $3.3 million in grants, of which the lion's share was used to build a youth and community center. An additional $1.2 million was granted to Homestead Habitat for Humanity to build two major housing projects.

Household disaster assistance

A critical component of community recovery is the extent to which individual households are able to recover. Household recovery – particularly rebuilding – depends upon two primary funding sources: private funds, such as household savings or insurance pay-outs, and federal and state public assistance programs. According to the South Dade Population Impact Study, in excess of 59 per cent of Florida City's households lacked insurance, compared to 47 per cent of Homestead's, and only 18 per cent of the remainder of unincorporated Dade County. While the number of households for which data were available for Florida City (18) and Homestead (89) was small, requiring cautious interpretation, it appears that insurance funds were a scarce resource, no doubt slowing the recovery process in both communities. A combination of socio-economic factors and availability issues, such as the red-lining discussed in Chapter 9, probably account for the low rate of insured homes. Regardless of the reasons, extensive alternative funding sources were required to facilitate the recovery of Florida City, as well as Homestead households.

Most of the recovery assistance available to individuals and households lacking sufficient insurance pay-outs or personal resources is administered through three federal and state programs. The first step in the assistance process is to apply for a Small Business Administration (SBA) low-interest loan, available to qualified households or businesses. For victims not qualifying for SBA loans or receiving insufficient SBA funds to rebuild and replace their losses, Individual Assistance (IA) and Individual Family Grants (IFG) programs are available. The IA program includes several federal initiatives, such as the provision of temporary housing. IFG is a federal (75 per cent) and state (25 per cent) co-funded program offering assistance to replace personal property not covered by insurance, such as home furnishings and clothing.

As a reflection of the level of poverty in Florida City, only 44 applicants or 5.5 per cent of its estimated homeowners qualified for SBA loans. By early 1993, Florida City residents had applied for a total of 637 IA and IFG grants. Of these, 221 IA and 277 IFG requests had been approved, representing a 78 per cent approval rate for a total of $1,526,371 in funds. While the approval rate is relatively high, the overall percentage of households getting funds was quite low. Based on an estimated 1,793 remaining households in Florida City (BEBR 1993), only 12.3 per cent received IA and 15.4 per cent received IFG monies.[2] This rate of assistance is quite low, when compared to Homestead. In Homestead, 715 SBA loans, or 20 per cent of its estimated homeowners, were approved. A total of 15,533 IA and IFG applications were processed. Of these, 6,051 IA and 7,632 IFG – or 88.1 per cent – were approved. Based on an estimated 9,357 households in Homestead, 65 per cent received IA grants and 82 per cent received IFG funds. These are considerable differences.

A major reason for the low proportion of Florida City households receiving IA/IFG funds was simply a much lower application rate. There was only one application for every 2.8 households in Florida City. In Homestead the rate was one for every 1.7 households. Why this difference? It is impossible to believe that Florida City's households did not need help – they were poorer, were even less likely to have insurance, and their average relative losses were greater. Expectations play a key role in motivating individuals to take action, so it is possible that many residents of Florida City did not apply because they perceived little chance of receiving help. For these mostly poor, minority families – many of whom are migratory laborers – the "system" simply did not work. As an illustration, when an elderly African-American woman whose insurance settlement was insufficient to cover her losses was asked by one of our researchers if she had applied to FEMA, she said no, explaining that she didn't think they would help. It is also likely that some immigrants in Florida City were uncomfortable about drawing attention to themselves and family members by approaching government authorities for aid. This uneasiness may be due to questionable documentation or simply to a historical distrust of government agencies. A third reason for the low application rate may have to do with confusion about the process. It took a considerable amount of savvy and perseverance to successfully work through the bureaucracy to obtain a grant. For example, it is difficult to comprehend that in order for a household to get federal assistance they had to first apply, and be turned down, for an SBA low-interest home loan. This continues to confuse disaster victims everywhere. A final reason is that poorer victims encounter more obstacles in making the necessary trips to the Disaster Assistance Centers due to transportation, child care, and work difficulties. Whatever the reasons, while the needs in Florida City were great, they were less likely to have received government assistance.

AN UNCERTAIN FUTURE

The picture that emerges for Florida City's full recovery from Hurricane Andrew is bleak. The city and its residents suffered extraordinary levels of damage, had limited access to insurance, and received low levels of public aid. Restoration and recovery has been much slower at both the community and household levels. There are some glimmers of hope, however. The plan to recreate Florida City into a "pioneer village" was abandoned and community leaders have acknowledged the importance of economic development and housing programs more in line with their constituents' needs. Several of the new housing programs are predicated on the goal of home ownership and include providing mortgage packages for low-income households. While these efforts are meeting with some success, many low-income renters will never be able to qualify, should they even want to buy a house. Some initiatives are addressing the rental problem, including the 166-unit complex approved to be built near Villages of Palm Bay (Hartman 1994). However, limitations have been placed on the project's density because of inadequate county sewer and fire/police services. Once again, external factors constrained attempts to meet community needs in Florida City.

One potential harbinger of economic recovery is the opening of a new $20 million outlet mall – Florida Keys Factory Shops. While its long-term economic success is uncertain, it increased Florida City's 1994 tax base by 30 per cent, bringing its total to nearly 87 per cent of its pre-Andrew level (Arthur 1994c). A new civic complex, including a city hall, is being funded by a $2.9 million grant from the Hurricane Andrew Recovery and Rebuilding Trust Fund, and at the time of writing it is nearing completion. Yet, these initiatives are not without their problems. With the opening of the factory outlet mall, it was discovered that building and permitting procedures had been violated. As a result, the county's code compliance division is now investigating all Florida City post-hurricane construction: commercial, public, and residential. The city acknowledges that difficulties in training new staff – in conjunction with pressures from residents and external interests to rebuild – may have resulted in compliance failures. Others have suggested that conflict of interest may have been involved (Arthur 1994d).

Despite these setbacks, past experience would indicate that, given sufficient time, community recovery will occur, or at least appear to occur, if assessed in terms of aggregate level indicators (Friesema *et al.* 1979; Wright *et al.* 1979). Given the level of destruction and considering its pre-impact status, Florida City need only show limited gains to surpass its pre-impact status. Nevertheless, there is a strong sense of opportunity missed. As one official expressed it, "After the hurricane, there existed an opportunity to build better and take advantage of potential." Recovery will undoubtedly be boosted by the fact that Florida City, and Homestead for that matter, is located in metropolitan Miami. One of the most diverse and fast-growing areas in the nation, Miami is squeezed between the coast to the east and the Everglades to the west. It is likely that Florida City will eventually be caught up in the southward urban expansion, resulting in further development.

If recovery eventually occurs, however, the question remains as to who will benefit: what will happen to the pre-Andrew residents of this small minority community? Will they recover? How many households will give up and leave the area? How many will remain, but will not have had a voice in community changes? Will the community be restored, or simply the area in which the community of Florida City once existed? These questions should be seriously addressed not only with respect to Florida City, but Homestead as well. While it appears that in the short run Homestead is faring better in the recovery process, only time will tell if the trend will hold. And even then, the above questions will remain relevant. The experiences of both communities can offer us valuable lessons for the future.

SOME POLICY CONSIDERATIONS

We can ask ourselves what could have facilitated the recovery of Florida City and its residents. Klinteberg (1979) and Rubin (1985) argue that the structure of community government is a critical factor. This case suggests that, while important, we need to look beyond internal factors to the ecological network in which recovery must occur. Within the larger political and social context of Dade County, this community has historically wielded minuscule power, and since Hurricane Andrew it has been unable to compete successfully for post-disaster resources, either public or private. And, at the household level, most residents have not been able to access sufficient resources for a timely recovery. In retrospect, these results are not surprising. Florida City and its residents were clearly a high disaster risk, not simply because of the location of their community in hurricane-prone South Florida, but because of their location in social space. A clear analysis of these risks demands that we examine more than geographical or physical attributes in our assessment of future impacted communities.

Mitigation efforts need to be focused on high-risk areas. This case study illustrates that these are not necessarily the locations most susceptible to hazardous events. Rather, they are communities left vulnerable because of social characteristics – both intrinsic and within the larger system in which they must operate. Any area with a high concentration of disenfranchised residents – people on the bottom of the power hierarchy, such as minorities and the poor – should be considered high risk. Households in these communities are unlikely to have the resources, human or material, to respond effectively to a natural disaster.

Specific demographic profiles can thus be used to predesignate target zones for special initiatives to assist these communities with information and resources with which to plan for disaster mitigation and recovery. Training is needed to help community leaders, from both the private and public sectors, prepare for disasters, including the development of recovery plans. Officials need special knowledge of aid qualification policies and grant procedures. Low-income housing should be targeted for mitigation assistance, such as grants and low-interest loans for shutters. A community-based center or door-to-door canvassing may be needed to reach those most in need of disaster-related grants and loans.

This case study has supported the assumption that effective local governmental response, such as was seen in the case of Homestead, is an important factor in successful disaster response. More importantly however, the case of Florida City suggests the argument that a community's position within the larger socio-political ecological network must also be considered a major factor in determining the extent to which it will be at risk for disproportionate impact, for slow recovery, and for disaster-driven community changes that are not likely to be in the best interests of its residents.

NOTES

1 Since the timing of measurements do not correspond with the disaster, some changes may not be exclusively due to its impact. Annual tax assessments are made in the first part of the year, initially released toward the end of summer, and the final figures released in October or November. Therefore, the 1992 assessment reflected property values prior to the storm and the 1993 assessment, while post-hurricane, may not have occurred long enough afterward to capture rebuilding and repair activities.

2 A given household could receive both Individual Assistance and Individual Family Grants funds, so some of these no doubt went to the same household, making the actual rate of households receiving any assistance even lower.

12

DISASTERS AND SOCIAL CHANGE

Hurricane Andrew and the reshaping of Miami?

Betty Hearn Morrow and Walter Gillis Peacock

The relationship between disasters and social change is an intriguing, albeit complex, one. There can be no denying that a natural disaster, particularly one of the magnitude of Hurricane Andrew, alters in dramatic fashion the bio-physical environment. The destruction and damage to homes, businesses, infrastructure, technology, and the natural environment required major alter-ations in "normal" activities – creating the disaster as a social event. Within a matter of hours Andrew qualitatively and quantitatively adjusted the nature and magnitude of needs of individuals, groups, and organizations throughout the Miami metropolitan area. Adaptation and change were required and inevitable.

For many of Miami's residents, while the natural environment will regrow and houses and businesses will be rebuilt, things will never be or feel the same. Their relationships and daily lives, indeed their sense of community, has been so altered that they have come to divide their lives into two segments – before and after Andrew.

By some accounts, initial responses to Andrew saw ethnic, racial, and class antagonisms seemingly dissipate, supplanted by community solidarity, unity, and the affirmation that despite its uniqueness, this was an American city and Americans pull together to face adversity (Portes and Stepick 1993). Not only did the community seemingly pull together, but the nation came to its aid. In a sense, post-Andrew Miami is different. Hurricane Andrew has provided its residents with a common history and discourse – a discourse that at least sym-bolically unites it and conjoins it to the nation as it had not been before. And yet, many of these apparent changes may prove to be short-lived.

Our work has employed a socio-political ecology perspective to focus on the social, economic, and political factors shaping the effects of Hurricane Andrew on the network of social systems that constitute the Miami metropolitan area. A central theme has been that while considerable alterations occurred as social units, from families to governments, attempted to regain their ability to function, to understand these alterations we cannot simply focus on the physical event itself nor treat the community as if it were a single integrated system. Rather, many of the social forces and structures shaping these alterations pre-dated the event. Furthermore, alterations and adjustments made in response to the event occur within a contested terrain in which access to resources, from information to capital, was uneven and not based exclusively upon need. As a consequence, some were better prepared, were able to marshal resources and move quickly along the road to recovery, perhaps even improving their pre-event status. For others, the process remains protracted and the results do not look promising. Our analyses have documented alterations in the network of social systems as new pressures and processes, as well as new groups and organizations, were brought to bear on problems exacerbated and created by the event. However, we have also seen remarkable continuity with preexisting patterns, particularly regarding the winners and losers in the recovery struggle. While change cannot be disputed, there is considerable uncertainty about its nature, rate, pervasiveness, duration and significance.

SOCIAL CHANGE AFTER DISASTER

A number of researchers suggest that, on the whole, disasters produce little in terms of consequential long-term change (Wright *et al.* 1979; Friesema *et al.* 1979; Rossi, Wright and Wright 1981). Others argue that disasters do not pro-mote innovative change, but simply accelerate or decelerate preexisting trends often grounded in historical cultural frameworks (Bates *et al.* 1963; Oliver-Smith 1986). Still others, following the lead of Samuel Prince (1920), suggest that disasters provide the impetus for major changes or alterations in the structures of impacted social systems (Dacy and Kunreuther 1969; Cochrane 1975; Abril-Ojeda 1982; Bates 1982; Killian, Peacock and Bates 1984; Hoover and Bates 1985).

There are undoubtedly a number of explanations for these seemingly inconsistent perspectives and findings – some inherent in the nature of social ecological networks. As we discussed at length in Chapter 2, communities are composed of potentially complex networks of interacting social units which must be considered when studying social change. The scale and complexity of a community's division of labor, its stratification systems and associated degree of inequality, and its political structures, must all be taken into account. While these are related dimensions, change in one is not necessarily accompanied by similar alterations in another. Cross-currents and divergent patterns are likely to emerge within each dimension, adding to the complexity, and to the impossibility of making a simple statement regarding the association between disasters and social change.

Other factors contributing to inconsistent research findings stem from differences in units of analysis and the nature of the event itself. Disaster researchers have studied everything from small, relatively isolated communities or a small number of households or organizations, to large metropolitan communities, clusters of communities and, in a few cases, entire societies. Results from one level of analysis may not be directly transferable to other levels. A more obvious difference among research findings stems from the events themselves, including factors such as their type, severity, and magnitude. Small-scale events are not as likely to engender change in overall community structure. Even after high-magnitude events, the potential for change is different for communities located in highly developed, post-industrial core societies, such as the case of Hurricane Andrew and Miami, than when they occur in peripheral or semi-peripheral societies.

Thus, when attempting to develop simplistic all-inclusive statements about the nature of social change after disasters, we are faced with a multitude of issues related to the complexity of social systems and networks, the variety and magnitude of disaster events, and the multi-dimensionality of change itself. And yet, despite the contradictions no one would deny that high-impact events destroy and damage the physical structures, technology, and natural environment – critical elements for the normal operation of social systems. In some situations these events result in a significant depopulation due to death and forced relocation. These modifications in the ecological field demand alterations or changes in the operation of the community's network. Rather than continuing to debate the issue, it is more productive to focus on the nature, duration, and extent of these changes.

Changes among the social units and networks can occur in the form of growth or decline, elaboration or simplification, and differentiation or consolidation (Bates and Peacock 1987). In this context it is important to consider, not only overall changes in a particular sector, but differential levels of change within it. It is likely that some parts of a community will experience growth, elaboration, and differentiation, while others will experience decline, simplification and consolidation, perhaps even disappear. As a result, it may be difficult to discern a single trend; rather, differential and contradictory patterns of change are likely to emerge, defying simple positive/negative categorization of their ultimate effects.

THE RESHAPING OF MIAMI?

In this concluding chapter we pull together evidence refuting and supporting the proposition that change resulted from Hurricane Andrew's impact on the socio-political ecology of Miami. We frame our discussion around arguments set forth by Bates and Peacock (1987) supporting the hypothesis that disasters provide impetus for major social changes. Using examples from the Guatemalan earthquake study (Bates 1982) to illustrate each proposition, Bates and Peacock ponder the relevance of each for other disaster impact settings where social systems may be organized and relate to larger systems in very different ways. The case of Hurricane Andrew in Miami provided an opportunity to assess the applicability of these arguments in a sharply contrasting social and political setting. The following summarizes the extent to which the evidence, most of which has been analyzed and discussed in previous chapters, supports or questions their six arguments suggesting that change may well follow in the wake of disaster.

Existing weaknesses revealed

Disasters place the social structure under stress, testing its capacity to perform vital functions. In the process, existing weaknesses are exacerbated and made visible. The system is forced to adapt and some of these adaptations will likely become permanent changes.

(Bates and Peacock 1987: 311)

In the case of Hurricane Andrew, community weaknesses during the warning and initial response phases were painfully obvious. Largely because it had been decades since a major hurricane hit the area, most organizations, including those with direct emergency management responsibilities, were poorly prepared. Shelters were dangerously inadequate and understaffed. In general, relations between the county Office of Emergency Management and other agencies and organizations were not well coordinated. Experiences associated with the near-miss of Hurricane Erin in 1995 suggested that many of the problems persisted (*Miami Herald* 1995, 23 August). Emergency management has since been reorganized and moved under the Metro Dade Fire Department. While this insures more resources brought under a single management structure, many of the problems of inter-agency and municipality coordination remain.

A serious weakness after Andrew was the general lack of effective local government response. In part this may be endemic to the metropolitan form of government. The fact that 90 per cent of the victims were not living within incorporated municipalities meant they lacked a neighborhood political entity to look to for direction and assistance, particularly in the early stages. Perhaps this could have been overcome by visible and effective leadership at the county level, but this did not materialize, except to the extent that particular individuals began to preform in leadership capacity on an ad hoc basis (Chapter 5). No strong political voice came forth during those first few days to represent the people of South Dade. The lines of responsibility were unclear at the local level.

Being incorporated did not guarantee attention and resources, as vividly illustrated by the case of Florida City (Chapter 11). However, municipalities with the organization and leadership in place to respond effectively, such as Homestead, drew immediate attention to their plight and were better able to marshal extensive resources, at least during the emergency and early recovery phases.

Election reforms took effect a few months after the storm and, instead of running county-wide, commissioners are now elected by district. As a consequence, minority political representation has increased dramatically. One result was increased attention to the long-term recovery needs and interests of neglected Black neighborhoods, as represented by initiatives such as the Moss Plan, a design for rebuilding South Dade led by Commissioner Dennis Moss, an African-American. In some communities, citizen movements for incorporation into new municipalities have subsequently emerged, several resulting in successful referenda. In response, Metro Dade established Team Metro offices in local neighborhoods throughout the county to increase its accessibility and presence. The county has now moved to a "strong mayor" system, which may provide more centralized leadership during the next crisis.

Absence of effective coordination and communication among the various levels of government became painfully obvious during the first few days, contributing to the delay in federal response. Accusations flew back and forth and several political figures took a great deal of heat. Hard questions continue to be asked, and have resulted in some changes in procedures and lines of authority among local, state, and federal officials.

There were no effective mechanisms in place to coordinate the efforts of various public and private social service agencies. Shortly after the hurricane, national staff from the American Red Cross helped establish a local VOLAG (Voluntary Agencies), an organization to coordinate the efforts of the voluntary agencies responding to Andrew. It later evolved into a very successful countywide VOAD (Voluntary Organizations Active in Disasters) which remains active. VOAD has undertaken a major incentive to help self-selected communities, such as Miami Springs, organize local Disaster Information and Response Centers (DIRCs) and enact local hurricane drills. An Unmet Needs Committee, initiated by the Red Cross and facilitated by Catholic Community Services, was extremely effective in coordinating efforts during the long recovery process to reach victims not being served through regular channels. Cooperating agencies were so convinced of its usefulness that it eventually moved beyond hurricane-related needs to become a permanent community coalition. The active participation by county government in many of these initiatives has improved the coordination between public and private agencies.

Market failures in several sectors, including housing and insurance, were revealed by Andrew. Its winds dramatically exposed many cases of faulty construction techniques, particularly in houses built since the 1960s. A series of articles in the local newspaper, followed by a grand jury investigation, documented all-too-cozy relationships between construction and development enterprises, local government officials, members of regulatory commissions, and building and zoning personnel (Getter 1992a; Getter 1992b). One result had been a steady reduction in the strength and enforcement of building codes, as well as the introduction of construction techniques, styles, and materials which were inappropriate for South Florida. As the reconstruction from Andrew began, the county appeared inept to deal with the monumental demand for building permits, inspections, and contractor regulation. As a result, many of these operations were decentralized into district offices. Building codes were revisited (Mitrani 1993), some components strengthened, and further changes are being considered. On the other hand, many reforms, such as banning mobile homes, were dropped primarily due to resistance from special-interest groups. The immediacy of rebuilding caused many mitigation initiatives to be put on hold. The extraordinary influence of the building industries continues to be a factor in local politics, bringing extreme pressure to bear on the county personnel and agencies responsible for building codes. It is apparent that while the pressure resulting from code failures has resulted in some reforms, the dynamics of the process have not changed. The ultimate result is likely to be that private economic interests will increasingly negate mitigation initiatives.

Perhaps one of the most significant economic consequences of Hurricane Andrew is the crisis it generated in the insurance market and industry (Collins Center for Public Policy 1995). The $15 billion payout by the insurance industry resulted in dramatic re-evaluations of exposure, and attempts by many companies to selectively withdraw from Florida's fire and casualty market. Were it not for a state-declared moratorium inhibiting partial withdrawal and a state-managed Joint Underwriting Association, homeowners' insurance might be unavailable today, bringing new development to a halt. Relatively few new private insurance policies have been written for Florida homes since 1992. The net effect is that the business of insurance and reinsurance – as well as who will pay when another disaster strikes – is radically different today than it was before Andrew. The implications for Florida, the nation, and the global insurance and reinsurance markets were there to be another disaster with losses as high as those with Andrew are potentially profound.

Andrew drew attention to the extent to which most of the Miami area lacked community-based social organization. This was largely the result of a long history of ineffective management of the development process. South Dade is a patchwork of poorly planned development with neighborhoods rarely existing as social entities. The few areas built as planned communities and with a history of activism, Country Walk for example, had a stronger voice in demanding representation and services from the county and in rallying against builders, insurance companies, or other factions standing in the way of their recovery. In those communities which came together to fight for their common interests, we would expect some residual empowerment and sense of community. On the other hand, depopulation, uneven reconstruction, and lack of consideration of pre-existing settlement patterns, for example in the FEMA trailer parks, reduced interaction or caused conflict in many neighborhoods, leaving them open for further neglect.

The social institution most sorely tested by this experience is the one most vital to recovery – the family. Even in non-disaster situations we are well aware that it is becoming increasingly difficult throughout society for families effectively to meet the physical and emotional needs of their members. The difficulties are, of course, exacerbated when human and material resources, including adults earning good wages, are lacking. Nowhere were the effects of preexisting resources more obviously related to recovery than at the household level. We have documented many ways in which gender, race, and ethnicity were associated with differential effects throughout this disaster experience. Carrying the analysis further, however, we have pointed out ways in which household structure, such as the ratio of wage-earners to dependents, the gender and age mix, and whether it fits the model being used by disaster responders, can profoundly affect a family's ability to recover. Many agencies and organizations discovered that their policies and procedures poorly fit the realities of South Dade households. Some subsequently adapted, albeit often late in the process. Hopefully, these adaptations will become institutionalized to reflect better the changing demographics and diversity of American families.

While this disaster sorely tested the capacity of many institutions to perform vital functions, the vastness of its impact and the involvement of large segments of the population in the painful and demanding process of recovery promoted the adoption of many stop-gap measures. Some reform has occurred and only time will tell if the rhetoric of looking for better ways to organize and mitigate against future disasters will result in permanent structural changes.

New groups and organizations

Disasters bring new groups and organizations into being and provide circumstances which foster new forms of contact, cooperation and conflict between existing groups and organizations.

(Bates and Peacock 1987: 312)

Many new groups emerged in South Florida to address various aspects of the recovery process, the most influential of which was We Will Rebuild. This umbrella organization of community leaders coordinated the distribution of nearly $28 million of private, and later public, funds (We Will Rebuild 1995). Building upon the model of Rebuild LA, the group has been acclaimed for its role in directing funds to a wide variety of needs, both large and small. As its funding and power grew, the organization was seen by many as being too focused on mainstream business and economic recovery. In response, it eventually became somewhat more inclusive in membership and more supportive of community infrastructure needs, such as daycare centers, recreation facilities, social services, and housing. It filled a particularly important niche in funding religious organizations that lacked resources and could not be assisted by the state and federal government. It would be hoped that any similar powerful group of elites emerging around future disasters or community crises will be more aware of the need to reach out actively to all segments of the community from the beginning.

One interesting result of the perception that the interests of women and children were not being adequately addressed by We Will Rebuild was the formation of a coalition of women's organizations into a counter-group. Women Will Rebuild set a new model for cooperative work among organizations from diverse segments of the community. While it disbanded when its hurricane-related goals were largely met, it is likely that these women's groups will continue to communicate and coalesce around common concerns. The coalitions formed in the creation of this organization may well have long-lasting implications for the Miami area.

Leaders from several Black areas formed a coalition to fight for recovery resources – the South Dade Alliance of leaders from Perrine, Goulds, Florida City, Princeton, and Richmond Heights – and the coalition continues to advocate for their interests. In a similar fashion, the Alliance for Aging was a major player in coordinating resources and assistance for elderly victims. Today it is actively involved in several projects to help older citizens mitigate against or be ready for future hurricanes. Centro Campesino has been a strong force for farmworkers' interests over the last two decades, providing sweat-equity home ownership opportunities (the practice of allowing buyers to build equity in their home by providing some of the construction labor), job training, and other community services. It was a focal point in relief activities, serving as a feeding station, base camp, medical and communication center for weeks after the storm. Its leadership aggressively secured grants and other funding incentives post-Andrew in order to accelerate its low-income housing programs. As a result, its resource base has been strengthened, spin-off organizations have been formed, and coalitions with other grass-roots organizations initiated. On a smaller scale, several other minority agencies, such as the Haitian Farmworkers' Association and the South Dade Immigration Service, have become more visible and active, expanding their mission to include many new social services.

The fact that so many poor people were left with damaged or destroyed homes tested the dedication and resources of the numerous national non-governmental organizations committed to helping communities rebuild after disasters. It also provided an opportunity for them to refine ways of coordinating their work. An interfaith coalition, ICARE, established a base camp for volunteer workers, developed a process with the county for regulating building permits for voluntary projects, and coordinated the assignments of dozens of church and non-profit groups. Its scale and effectiveness appear to have set a new precedent and will, no doubt, make the task easier the next time. On the other hand, some groups preferred to remain autonomous, such as the Salvation Army which established its own base camp and coalitions.

Hurricane Andrew drew national attention to the acute need for housing for low-income families. As a result, Habitat for Humanity expanded its presence in the area to several major projects, including a planned 200-home model community. Similar projects promoting home ownership are in various stages of development, sponsored by a myriad of public and private sources. While these projects meet an important need, not every poor family can afford to, or even desires to, own a home. Similar efforts related to low-income rental housing

233

were slower. Indeed, there has been a reluctance even to rebuild the grossly inadequate stock of subsidized apartments.

Some national groups active in disasters redefined their missions as a result of the extensive recovery needs encountered in South Florida. The American Red Cross established an unprecedented long-term recovery project in hopes of developing a model for future events of this magnitude. While the Red Cross is not likely to duplicate its Andrew recovery project, it has identified ways in which it can help other agencies through training programs in case management, as one example.

The FEMA Hurricane Andrew response was its most extensive, and expensive, project to date. The enormity of the needs, as well as the cultural diversity of the victims, presented many new challenges to the organization. While we have pointed out some of the more obvious shortcomings in the initial FEMA response, in all fairness there are many examples where procedures, even policies, were changed to adapt to the realities of local conditions, such as in the case of the "one head of household" rule which was later somewhat modified to account for multifamily households. One factor which complicated FEMA's work was the sometimes conflicting information and recommendations they received from various governmental levels, including the Presidential Task Force, as well as between different levels and sectors within their own organization. FEMA managers were sometimes faced with contradictory directives, or "second-guessed" by administrators who were not familiar with the field conditions they were encountering. Some of this was likely political, as in the case of being directed to "fast track" as many applications as possible (Polny 1993) (a Presidential campaign was going on) even though there was no way for victims to handle or use the money, i.e. there were no stores or regular banks operating in South Dade during those first weeks. As a consequence, much of that early money was inappropriately distributed or improperly used by recipients. If our more recent observations in the US Virgin Islands during the 1995 Hurricane Marilyn response is any example (Morrow and Ragsdale 1996), FEMA learned many valuable lessons from the Hurricane Andrew experience and has significantly altered many of its procedures.

Immediately after the storm, in communities and neighborhoods where authorities were slow to respond or non-existent, local churches often took the initiative to coordinate relief efforts. In communities such as South Miami Heights, Perrine, and Goulds, neighborhood churches remain a major source of assistance for families continuing to experience difficulties. Many church groups applied for and received grants to expand their mission to include new services, such as mental health counseling. Prior to Andrew, few social service agencies had offices in the southern part of the county, in spite of the high number of needy residents living there. Today most do, and many, such as Catholic Community Services and Lutheran Ministries of Florida, have established permanent offices in the area.

External resources and ideas

> Disasters frequently result in a large influx of outside resources, both human and material. This may produce an economic boom, as well as bring in new ideas and ways of doing things. These outside resources, ideas, and behaviors can result in fundamental changes in the community and its social structure.
>
> (Bates and Peacock 1987: 312)

The influx of capital and physical resources, labor, new information and ways of doing things have strong implications for change. While few would deny the necessity of external resources, the inundation by outsiders can be overwhelming, sometimes almost creating a second disaster.

Unlike some Western industrialized nations where the national government takes a direct role in financing, planning, and implementing the reconstruction process, the United States relies heavily upon private insurance payments and the actions of voluntary agencies, supplemented by government-sponsored low-interest loans and grants. There is no overarching integrated organizational plan, but rather a market-based recovery approach is supplemented by indirect governmental activities.

In the case of Andrew, there were some exceptions. The State of Florida enjoyed a phenomenal increase in sales tax collections due to the purchase of an unprecedented volume of construction materials and household replacement items. The legislature set a large proportion of this money aside in a Hurricane Trust Fund. Monies from this trust were filtered through We Will Rebuild and other agencies to fund a host of different projects, including non-profit community-based housing developments. These funds served to promote a new type of community-based organization that not only provides housing for those not otherwise able to afford it, but also educational, legal, and other empowerment services.

The largest infusion of capital into the area was from insurance pay-outs. The sales of building materials, services, furnishings, automobiles, and other replacement products created a tremendous retail boom in South Florida. Many new businesses, such as carpet, furniture, and building supply stores, opened to meet the demands. A wholesale outlet mall was built in Florida City. While these new enterprises were a welcome sign of faith in the community's recovery, it remains to be seen whether they will be able to maintain sufficient sales volume to be profitable in the long term.

National and international corporations, with their outside resources, were able to open, or reopen, outlets and franchises before smaller, independently owned stores could negotiate the insurance and reconstruction processes. As an example, the first businesses to open in the Homestead area were fast-food chains, such as Checkers and McDonalds. Small local businesses have always been a vital part of Miami's economy, particularly among immigrant populations, but, even before Andrew, they were finding it harder and harder to compete. It appears that the hurricane has accelerated this economic restructuring away from locally owned to national and multinational firms.

Extra-local labor entered the Miami area in unprecedented numbers as construction workers came in search of work. Most were single males, but entire households came as well, further exacerbating the housing crisis. Workers lived in the shells of damaged houses, trailers, tents, and trucks throughout the area. While many were law-abiding and worked hard to help rebuild the area, others created a range of problems, including swindling homeowners, public and domestic violence, burglary, and other crimes and nuisances. The influence of these "roofers from hell," as some were called, on impacted communities is well known, such as the case of the island of St. Croix after Hurricane Hugo, where their presence is thought to have permanently changed the culture, particularly in relation to illicit drugs (Morrow 1992). Thousands of volunteers came to South Florida as representatives of agencies, churches, and colleges to help with the relief and reconstruction, many staying for months or returning several times. Their contributions in terms of moral support, as well as labor, were enormous and, in many cases, individual and organizational connections continue. Some of the novel ideas related to post-disaster redevelopment are likely to have come from these individuals and many have remained as permanent employees and directors of non-governmental agency programs.

No doubt many local residents and organizations are different today as a result of the influence of these outsiders, both good and bad. In South Florida, in contrast to a disaster situation in a developing country or small rural community, however, we are dealing with a diverse population, even by US standards, which is constantly exposed to new ideas and behavior as immigrants, new residents, and tourists come into the area. The extent to which new or innovative information was introduced as a result of this disaster is difficult to gauge. Post-industrial societies are based to a large extent on the processing, storage, and transmission of information. It is likely that much of what would be considered new information was already here, though perhaps not in an available location in the social network where it could be utilized.

One example of a technology transfer accelerated by the hurricane was the use of geographical information systems (GIS) in various mapping tasks, such as damage assessment and recovery monitoring. A focal point in the use of GIS was FEMA's project to develop an integrated data system to guide the rebuilding process. This approach stimulated the development of similar GIS capabilities in the Metro Dade Planning, Development and Regulation Department. Several conferences were held with the purpose of introducing new information and expertise into the community, such as two sponsored by the Knight Foundation and the University of Miami where disaster experts were brought in to discuss all facets of disaster relief and reconstruction. The extent to which this information was new, was incorporated into the local network of systems, or stimulated change is difficult to assess.

Some new approaches to community planning, such as the *charrette* concept, were introduced. Charrettes are intensive, concentrated work sessions of architects, urban planners, and others which are designed to produce workable creative projects in a short period of time. Following Andrew, several charrettes were held and plans developed to guide the rebuilding of South Dade as a whole, as well as selected areas. While emphasizing economic revitalization, they also encouraged the development of small parks and community centers, also usable as shelters, serving as the social focus for neighborhoods. Some attempts to impose new house designs and housing development patterns were met with resentment, particularly by local ethnic communities. Despite intentions to be culturally sensitive, planners sometimes failed to consult local groups and, therefore, made assumptions which failed to account for local aspirations and goals. Without sufficient political or economic power behind them, however, these ideas have been slow to be implemented or incorporated into actual building projects.

In summary, human and material resources poured into Dade County after Hurricane Andrew. As far as information and technologies, it is doubtful that ideas were introduced which would not eventually have reached Miami, but no doubt there are cases of accelerated dissemination, and in the case of some redevelopment ideas may have been very slow to appear. With a few notable exceptions, most material resources were directed to replacing what was already there or meeting short-term relief needs. Nearly all of the southern part of this metropolitan area has been rebuilt, much of it better than before. There is no question that, on the whole, the economy of South Dade benefited from hurricane-related resources. The extent to which these benefits are long-term and sustainable has yet to be determined. And, we may well ask, who has benefited?

Differential effects on preexisting strata

Disasters differentially affect socio-economic and ethnic groups, as well as different sectors of the community's division of labor. As a consequence the stratification system may be affected and differential decline and growth may occur in various sectors of the social structure.

(Bates and Peacock 1987: 311–12)

A central theme in our work has been ways in which the hurricane-related experiences of ethnic and racial minorities, or their communities, differed. Minority households, particularly Black households, reported significantly greater amounts of damage from the storm. Our data also suggest clear and statistically significant differences in insurance coverage. This was another weakness exposed by Hurricane Andrew. In light of the greater damage and the critical role of insurance, these households are at higher risk of not recovering from Andrew. Most disturbing are the differences which appeared when we asked

237

victims with homeowners' insurance if their insurance settlements were sufficient to cover their rebuilding and reconstruction expenses. Black and non-Cuban Hispanic households, across income levels, were much more likely than Anglo and Cuban households to report insufficient settlements and this was in part due to differential access to policies with larger corporations. The red-lining and equity issues behind these findings are not being addressed by the state in its attempts to deal with the insurance crisis.

The occurrence of differential access to insurance and reduced settlements coupled with racial segregation is likely to result in concentrated areas of lagging recovery. Furthermore, the case study of Florida City shows ways in which entire minority communities can be left behind in disaster recovery. The net result is a high probability that inter-ethnic inequality in the Miami area increased.

Inequality of recovery may well contribute to alterations in the spatial distribution of ethnic groups. Over the last several decades, as Dade County became increasingly Hispanic, many Anglo households moved north to Broward County. Shortly after the hurricane, increased 'White flight' was predicted as Anglo households seized this opportunity to use their insurance money to pay off their mortgages, sell their homes, and move north. As suggested in Chapter 10, the issue of accelerated White flight remains open, but it is clear that Black households experienced extraordinary difficulty in relocating following Andrew. To the extent that these difficulties reflect continued discrimination and marginalization in housing, the Miami area is likely to continue to have relatively high levels of Black residential segregation.

Many landlords and developers hoped to increase profits by taking advantage of the opportunity to upgrade – replacing apartment and commercial rentals with more expensive units. Some communities discouraged or slowed down the rebuilding of low-income housing, hoping that poor and/or minority residents would relocate. There has been an obvious hesitation to replace mobile home parks. While this may be rational from a mitigation standpoint, it reduces an important source of low-cost housing, particularly for retired elderly. In total, it is probable that some areas of rebuilt South Dade will have fewer low-income families, and they will be concentrated into fewer areas.

We have already discussed several initiatives to build low-income homes. While slow to happen, eventually most of the low-income housing, rental as well as owner-occupied, was replaced. There is some speculation that, at least in deepest South Dade, the upper and lower classes have largely remained and recovered, but much of the middle class has left. This suggests a fruitful topic for further investigation.

A pioneering element in our analysis has been its emphasis on ways in which gender impacts disaster experiences, both for victims and responders. Women were heavily represented among those providing services to victims, but less so in key leadership roles. A few exceptions, however, may have served to open some doors for the next disaster. Female victims, particularly those without partners and charged with the care of children or elderly parents, were especially burdened with extra work and responsibilities. An expected finding was that households with few resources going into this disaster experienced greater recovery problems. Putting all of this together, the households left behind in the recovery were disproportionately headed by poor minority women. We see no signs of greater institutional attention to their plight. Indeed, the tide appears to be going the other way.

At the family level, problems associated with Andrew have led to tremendous internal stress, sometimes conflict. The extent to which this resulted in families breaking up is unclear, but there is reason to believe that the roles and relationships within many families were permanently changed as a result of going through this long ordeal.

At the agency level, we have illustrated many ways in which response organizations, public and private, were not prepared to deal with the diversity of Miami's population. Extensive drawn-out application processes showed little concern for the difficulties encountered by applicants who were poor, spoke a different language, lacked transportation and child care, were elderly or in poor health, or lived in non-standard households. As the problems became obvious, many agencies made adjustments and, hopefully, the Hurricane Andrew experience will have resulted in some permanent improvements in assistance policies and procedures.

In sum it is clear that, as with other disasters, the poorer and more disadvantaged suffered disproportionately and had greater difficulty recovering, a pattern clearly consistent with more "normal" social processes. It is likely that the consequences will be greater inequality and heightened segregation. However, countervailing these expectations is increased mobilization of these communities through voluntary and non-governmental organizations, and better political representation.

Changed infrastructure

Disasters frequently destroy or severely damage outmoded infrastructure and force its replacement by more modern technology. Such technological innovation may result in alteration of the stratification system or the division of labor, and may result in both differential growth and elaboration of sectors of the system's structure.

(Bates and Peacock 1987: 312)

This particular catalyst of change is not as applicable to our case as it would be to a developing nation or older community. Because the Miami area is relatively young, most of its infrastructure is only a few decades old, at most. Yet, the destruction presented some clear opportunities to update or replace outmoded technology, such as electric and telephone services. Over 3,000 miles of poles were knocked down by the storm, creating millions of dollars of damage and leaving most of the area without service for weeks and months. In those few areas where the lines were buried, little damage occurred and service resumed within a few days. An obvious choice would have been to rebuild the power and telephone systems using buried lines. However, pressure to get service restored prevented serious consideration of this option. Interestingly, the decision was also justified on the basis of a cost-benefit analysis which compared repair costs of in-ground and above-ground lines from 1985 through 1991 – years with no hurricanes.

New radar, communication, and computer technologies were introduced throughout Dade, either in response to damage to older equipment or its perceived inadequacy. The destruction of Homestead Air Force Base hastened its demise as a major facility and caused considerable demographic change as thousands of military and support personnel permanently left the area. After the community mounted a major appeal, it is being rebuilt on a much smaller scale, but with new, state-of-the-art equipment. Some of the National Hurricane Center's equipment was damaged. The Center now has new antennae, updated equipment, and has moved into new more modern and wind-resistant facilities on the campus of Florida International University. Some office buildings, including the headquarters of American Banking Association and the Burger King Corporation, were rebuilt with special attention to mitigation, including the protection of computer and communications systems. The new hospital in Homestead has undergone major improvements in equipment and facilities, as is true of many of the facilities serving the people of South Dade.

The storm revealed many construction weaknesses and experts subsequently suggested many ways in which houses could be built to withstand higher winds. Largely as a result of arguments by the building industry that new requirements would increase costs and reduce the availability of housing, many of these reforms were rejected. There are new minimum standards, however, for windows, doors, roofing procedures, shingles, and shutters. Most of the new low-income homes, such as those being built by Habitat for Humanity and Centro Campesino, are designed to be hurricane-resistant. However, most rebuilding occurred with only slight mitigation modifications, often using older building codes, and an unknown amount was illegally completed without any permits or inspections. The incorporation of new technologies in overall building is likely to be much less than hoped in the short run.

Conflict over scarce resources

Conflicts often emerge in the aftermath of a disaster over the distribution of scarce resources and over the equity principles which should guide the reconstruction effort. These conflicts may have serious political implications and result in permanent changes in the relationships between the government and other units comprising the system.

(Bates and Peacock 1987: 312)

240

Recovery, demanding as it does the allocation and reallocation of scarce resources, is inherently a conflict-ridden process. Friction is endemic to all asymmetrical relationships as subordinate elements within the social network seek to improve their lot and the powerful act to maintain their position. The tough decisions over finite reconstruction resources force communities to deal with basic issues of equity, participation, and inclusiveness.

Perhaps most interesting is the extent to which equity principles are not addressed. As we have discussed, recovery in the United States is based predominantly on the market and rarely are inequities associated with the market questioned. Such was the case following Hurricane Andrew. Issues related to differential access to housing and insurance, particularly with respect to African-Americans, have remained largely in the background. As long as these assumptions remain unquestioned, the possibility of correcting the issues is nonexistent. And yet, with respect to public dollars and funding, the discourse on equity is far from silent.

We have already noted the political mobilization of new groups, such as Women Will Rebuild and the South Dade Alliance, which struggled to have a voice in the recovery process. These groups represent departures from the recent past – their members came from segments of the population with little or no tradition of working together for common causes. There was considerable friction about whether any special disaster and redevelopment funds should be used outside the heavily impacted area. Federal authorities suggested an 85 per cent to 15 per cent split which proved to be very controversial. As examples, the decision to use some We Will Rebuild funds to help revitalize a shopping plaza in downtown Miami was met with anger and strong resistance by residents of South Dade. Also, an unprecedented coalition among Hispanic, Haitian, and Black leaders in Homestead has formed to question certain urban redevelopment projects. While past decisions like this would probably have gone unchallenged, the hurricane left many groups more vocal and politically active. And, with the new district voting patterns, they now have a better means for applying pressure on local representatives.

New groups and coalitions emerged and evolved in response to the needs of Andrew victims and the opportunities associated with the influx of resources. Most followed well-established patterns of structure and power. However, there were instances in which minorities and formerly disenfranchised groups became vitalized and effectively directed resources toward the under-served. This empowerment of women, the elderly, and minorities is likely to have some lasting effects on the establishment, at least in terms of awareness and lip service, if not real power-sharing.

SUMMARY

We began this chapter acknowledging the complexities of trying to develop a single statement about the direction or nature of change following a natural disaster. We noted the potential for conflicting and/or contradictory change within and between sectors of a community. In many ways our ensuing discussion of Bates and Peacock's (1987) arguments supporting the expectation of social change has demonstrated the expected complexities and contradictions.

Weaknesses within the social system were exposed and in many cases addressed, at least in the short term. The inherent conflictual nature of the recovery process resulted in the formation of new alliances and groups to challenge the status quo. Indeed, the success of these challenges is reflected in the inclusiveness and diversity within the public and private groups coordinating community redevelopment projects. Nevertheless, evidence in terms of household recovery clearly suggests the potential for increased ethnic inequality and spatial segregation, or at least little in the way of a reversal of trends. It is likely, given the economic dynamics of the Miami metropolitan area, that overall changes will be registered when examining the aggregate level numbers for the area; however, the consequences for South Dade and for local business interest is unclear. And the slow pace of local business recovery has long-term consequences for the economic future of at least segments of the community.

Thus, while some changes reflect alterations in the nature of this complex multiethnic community, others reflect a reversion to or at least the maintenance of the status quo, perhaps even exacerbating preexisting stratification patterns. The overall reshaping of Miami, remains an open question that only time will answer. At some levels and in certain sectors Miami has been reshaped, while other potential changes are being thwarted, and in still others things happen as if Andrew never occurred. Will Miami be ready for the next storm or will the community that emerged from Andrew be as ill-prepared both in terms of its built environment and its social infrastructure as the one that faced Andrew? If the really big one hits – the one that comes on shore over the densely populated beach areas and then plows through the downtown financial and business districts – will Miami, the state, and the nation be ready? In light of the fact that South Florida reflects the future demographics of the United States, particularly those regions subject to high impact disasters, important lessons can and must be learned from a detailed tracking and re-examination of the directions and decisions taken by this complex urban ecological network.

APPENDIX
Hurricane Andrew research projects

TENT CITY STUDY

Walter Gillis Peacock (Principal Investigator). Funding: Office of Sponsored Research, Florida International University.

Our first research activity consisted of interviewing in the relief centers or "tent cities" established in and around Homestead and Florida City as temporary shelters for those left homeless by the storm. The team conducted open-ended, in-depth interviews with people living in two of the centers, as well as personnel from the American Red Cross, Federal Emergency Management Agency, and the various military units in charge. We interviewed over fifty individuals and households. Two team members (Yelvington and Kerner) stayed overnight there to gain a better understanding of the daily routines and problems encountered by residents and managers.

These interviews provided information about the problems and experiences of severely impacted victims left with no other place to live, formulating many of the questions which guided our subsequent work. By studying their organization, management, and internal relationships, we gained insight into: (1) the function of these temporary relief centers in disaster response; (2) the larger social forces that determined who ended up there; and (3) what the experience was like for various segments of the tent city communities. About one-third of the interviews were conducted in Spanish and then translated. All were transcribed and coded around emergent themes. Respondents were not drawn from a random sample, but were approached by team members in the course of participant-observation fieldwork. In total, members of the team visited the tent cities two or three times per week for several weeks, spending three to five hours there each visit.

FIU HURRICANE ANDREW SURVEY

Walter Gillis Peacock (Principal Investigator) and Hugh Gladwin (Co-Principal Investigator). Funding: National Science Foundation Grant #9224537; John S. and James L. Knight Foundation Grant #581801000.

The telephone survey of over 1,600 households was conducted using random digit dialing techniques. The survey's primary focus was households in the Miami metropolitan area (Dade County) and a smaller sample of 300 was obtained in Broward County, just to the north of Miami. The Dade County sample includes a disproportionate sample of 504 households that lived in the high impact zone of South Dade (south of S.W. 88th Street, also known as Kendall Drive) before Hurricane Andrew (identified by respondents' ZIP Code before the hurricane). Of these 504 former South Dade Households, 484 represent an over-sample reached initially through dialing randomly selected South Dade telephone exchanges. Households that were displaced were pursued via call-forwarding, new numbers issued by the telephone company, or answering machine messages. The over-sample includes the random sample for all of Dade County, plus an additional 300 households contacted by dialing randomly generated numbers with South Dade telephone exchanges and asking if the household lived South of Kendall before Hurricane Andrew. In addition to the over-sample, 20 former South Dade households were contacted when calling telephone exchanges in the initial random sample outside of South Dade. The Dade sample is weighted to reflect better the entire population of Dade County. The 132-question survey instrument asked about household preparation activities, evacuation, household damage and repair, displacement, insurance, other forms of aid and assistance, problems related to recovery, as well as their future plans. Extensive demographic information was collected on each household. Where called for, the interviews were conducted in Spanish.

FAMILY IMPACT STUDY

Betty Hearn Morrow and Elaine Enarson, Co-Investigators.

Andrew destroyed homes, not just houses. Disruption in South Dade took many different forms, including household and job loss or dislocation, disrupted commuting patterns, living in crowded and often deteriorated structures, the maze of paperwork and tasks associated with loss recovery and household reconstruction, and the lack of community infrastructure – parks, recreation facilities, neighborhood stores, and services. Morrow and Enarson were interested in how this crisis was affecting families, particularly in those segments of the community where recovery seemed to be progressing more slowly. Women were used as primary informants since they were more likely to be responsible for day-to-day family activities. Open-ended interviews were held with a purposive sample of victims and care providers in Dade County – often one person played both roles. Focus groups were conducted with several groups identified by agency caseworkers as having a particularly difficult time, such as single mothers in public housing projects, low-income Haitian women, family day care providers, and residents of a shelter for battered women. Represented in this project were the American Red Cross, FEMA, Alliance for Aging, Young Women's Christian Association, VISTA, Catholic Services, Salvation Army, Lutheran Ministries, National Association of Women's Business Organizations, Florida Health and Rehabilitation Services, Interfaith Coalition, Christian Community Service Agency, United Methodist Disaster Response, Metropolitan Dade County Emergency Management, Coalition of Farmworkers' Associations, Centro Campesino, South Dade Immigration Services, Haitian Women's Coalition, Switchboard of Miami, Women and Children First, Legal Services of Miami, Women's Emergency Network, SafeSpace, Parent Resource Center, and the South Dade Family Coalition. All interviews and focus groups were transcribed and, when necessary, translated from Spanish to English.

APPENDIX

AMERICAN RED CROSS PROJECT

Betty Hearn Morrow, (Director), Walter Gillis Peacock (Co-Director), and Elaine Enarson (Faculty Associate). Funded by the American Red Cross Hurricane Andrew Recovery Project.

In a study conducted for the American Red Cross (ARC), its unprecedented Hurricane Andrew Recovery Project was evaluated as part of an on-going national effort to determine the most effective recovery function for the ARC after a high-impact disaster. In-depth interviews were conducted with about fifty key informants from the Red Cross and other agencies involved in the South Florida recovery effort. All interviews were transcribed, coded, and sorted using the Ethnograph software program. Morrow, Enarson, and Peacock then identified several areas in which the ARC could be particularly helpful in long-term recovery. The ARC's long-term client intervention in South Dade met with limited success and was the subject of considerable controversy within ARC and the community. It was recommended that in the future, when there are many agencies working in a stricken area, the ARC considers providing its long-term assistance indirectly through other agencies and focuses its attention on providing leadership and training in client case management to these providers. The organization's unique relationship with FEMA places it in a central position to serve as a clearinghouse for client information.

SOUTH MIAMI HEIGHTS SURVEY

Walter Gillis Peacock and Betty Hearn Morrow (Co-Directors). Funded as part of National Science Foundation Grant #9224537.

To obtain detailed information about household reconstruction and recovery, the team focused on one neighborhood and completed an intensive door-to-door study. South Miami Heights, an unincorporated neighborhood of mostly single-family homes, was purposively selected. This community of primarily modest, single-family homes was heavily damaged. According to 1990 Census data, it was a culturally diverse community – 49 per cent Hispanic, 28 per cent Black, and 23 per cent Anglo – consisting mostly of working class (middle to lower-middle) homeowners. It was heavily impacted by the hurricane, yet virtually ignored by the military, Red Cross, and other agencies.

Extensive damage to housing in the area required that detailed mapping be completed before a sample could be drawn. Initially Census blocks were mapped out on aerial photographs of the area, with larger blocks being subdivided into more standard blocks including approximately twenty housing units. All blocks containing multi-family units were excluded because all of these units were destroyed or so heavily damaged that they were virtually abandoned. In addition, two rural blocks containing very few housing units were excluded. On the basis of this mapping, 377 blocks were identified, containing on average 19 housing units. From these, 99 blocks were randomly selected for detailed mapping. The detailed mapping identified 1,879 dwelling units, of which 1,501 were occupied and constituted the sampling frame. A random sample was generated and 261 interviews completed. Structured interviews were conducted with each of these households using a questionnaire which contained eighty questions related to sources and amounts of assistance received, insurance, contractors, evaluations of community recovery, problems and needs, and the impact of Andrew and the recovery process on household members. In addition, interviewers were asked to record their impressions and comments, resulting in some important ethnographic data.

SOUTH DADE POPULATION IMPACT STUDY

Hugh Gladwin, Walter Gillis Peacock, and Chris Girard (Co-Directors). Funding: State of Florida; Bureau of Economic and Business Research, University of Florida; supplemented by National Science Foundation Grant #9224537.

Each year the Bureau of Economic and Business Research at the University of Florida derives estimates of population change for the State of Florida to use in allocating funds and services. They normally use tax assessor and building permit totals to estimate the stock of housing, which is then combined with information on average household size to project the Census estimates to the current year. This procedure would almost certainly have led to a major undercount in South Dade in 1993, since the stock of housing had been drastically reduced and it was quite probable that household size had been altered as families doubled and sometimes tripled up. Furthermore, there were likely to be higher rental densities and larger numbers of people living in non-standard arrangements. In light of these potential shortcomings, the Bureau of Economic and Business Research sub-contracted with FIU researchers to assist in obtaining a more accurate estimate of Dade's population.

To increase precision further, particularly regarding estimates of occupancy rates, The data were collected using a sample designed to achieve the twin goals of estimating post-Hurricane Andrew occupancy rates and determining the number of persons per housing unit in the southern area of Dade County, Homestead, and Florida City. A stratified cluster sample was employed to achieve maximum precision. The sampling frame comprised 4,763 Census blocks in Dade County south of North Kendall Drive, and was divided initially into three strata based on number of housing units and per cent rentals in the 1990 Census. The first stratum consists of 812 "marginal" blocks which had five or fewer housing units in each block. According to the 1990 Census, these blocks contained only 2,243 units, comprising less than 2 per cent of the 127,635 housing units in the southern part of Dade County. The second stratum included 69 "special" blocks, which contained either more than 300 housing units or more than 95 per cent rentals and more than 50 housing units. Special blocks contained a total of 32,872 housing units, or about one-quarter of housing units in the southern part of Dade County in 1990. The third stratum of "regular" blocks comprised the remaining blocks, holding 92,520 housing units in 1990.

To increase precision further, particularly regarding estimates of occupancy rates, maps of the route of Hurricane Andrew (published in the *Miami Herald*) were used to create impact areas within regular and special blocks. Area 1, roughly located between S.W. 88th (Kendall) and S.W. 104th Street (Zone 3 in Figure 9.1, page 177), was largely outside the eye wall. Areas 2 and 3 (which together make up Zone 2 in Figure 9.1) are horizontal bands just south of Area 1 and were in the path of Hurricane Andrew's eye wall. Area 2 extends down to the region around S.W. 136th Street and Area 3 extends from there to S.W. 200th Street. This north eye wall was split into two zones primarily because Metro Dade estimates of loss of housing units

246

from the hurricane indicated substantial differences in the percentage housing loss in the northern and southern portions of this area (Metro Dade Planning Department 1993). Area 4 (Zone 1 in Figure 9.1), which is below S.W. 200th, was in the path of Hurricane Andrew's eye and sustained the highest damage. Within this area, blocks in Homestead and Florida City were separated to create Areas 5 and 6 respectively, so that separate estimates could be made for each of these two incorporated towns.

Within each impact area, regular blocks were randomly selected using a probability proportional to size (PPS) method (Kish 1965). The number of blocks selected within each area was determined initially by the number of housing units in that zone according to the 1990 Census. Then, the number to be sampled was adjusted slightly upward if damage estimates were high. For each of the areas, the probability of selecting a housing unit was (roughly) as follows: Areas 1 and 2: 1.6 per cent; Areas 3 and 4: 1.8 per cent; Area 5: 5.3 per cent; and Area 6: 13.3 per cent.

Once selected for inclusion, the housing units in each block were remapped by teams who supplemented their mapping activities with real estate maps, aerial photos, and tax portfolio data. After detailed maps had been created, randomly generated numbers were used to select housing units within blocks. Ten housing units were sampled in each selected block with ten or more housing units in 1990 and for which the number of housing units counted by our interviewers equaled the 1990 Census count. In blocks with six to ten housing units in the 1990 Census, all housing units in the block were selected for interviewing. In total 218 regular blocks were selected, remapped, and their housing units subsequently sampled.

Marginal blocks, constituting less than 2 per cent of South Dade's housing units in 1990, were eliminated from the sampling frame to increase the efficiency of estimates at the expense of minimal bias. While only 69 of the 4,763 Census blocks were designated as special blocks, they required careful attention because their housing units comprise approximately one-quarter of the housing units in South Dade. The need for careful treatment was further accentuated by the "lumpiness" of hurricane damage and subsequent rebuilding. Specifically, hurricane damage tended to occur in clusters and pockets, due to a variety of factors including building construction techniques used in particular housing developments and apartment complexes and variability within the storm itself. The problem is especially acute in blocks with large multiple-dwelling complexes, because apartments and condominiums tend to be repaired or left abandoned as a group. With so few special blocks and wide variation in block size (one block in West Kendall had more than 2,000 housing units), PPS selection would have produced considerable variance from sample to sample. As a result, special blocks within Areas 1 to 4 were selected using simple random sampling. Then, housing units within special blocks were selected using systematic sampling (1 out of every 19 units). As suggested by Kish (1965), running totals of the count were kept and carried over from block to block, to eliminate the problem of fractions (most blocks do not have a total count that is a multiple of nineteen). In Homestead, all eight of the special blocks identified were sampled. In Florida City, no special blocks were identified. Out of a total of thirty special blocks selected, two, one of which was in Homestead, had to be thrown out because the Census mistakenly assigned housing units to blocks that never had any housing units. These blocks were either empty or had only a church and related buildings (confirmed by 1992 aerial photos taken before Hurricane Andrew).

It must be noted that to determine the actual number of housing units sampled, the structures within each block were mapped to ascertain the actual number of dwelling units located in the block. All existing housing units were added to any housing units that had been destroyed by Hurricane Andrew and had not been replaced. Developers, neighbors, county tax data, and aerial photos were consulted to identify all vacant lots that may have contained structures having housing units before Hurricane Andrew, and hence should be included in our sample, but were subsequently cleared of destroyed structures. New units that replaced those destroyed by the hurricane were counted as if they were the old units under repair, and, therefore, included in the survey. New houses (built since Hurricane Andrew) were not counted as housing units if outward appearance clearly indicated they were still under construction and not yet ready for occupancy. In cases in which the actual count of housing units in our survey is smaller than the 1990 Census count, the reason is an error made by the Census rather than the loss of units.

Substantial differences between our count and the Census count are fairly common, particularly the further south the Census block was. Although new construction accounts for some of these differences, in most cases the Census count is simply wrong. Tax roll data and aerial photographs were frequently used to confirm our count. Not infrequently, the Census would double-count housing units that were in a block within another block, or assign units to the wrong block. More often, the Census count was simply too low, judging by the age of the buildings/structures in existence and the fact that tax rolls showed nothing new in the block since 1990.

In South Dade, 248 Census blocks (218 regular and 30 special) were selected, mapped, and sampled, resulting in 2,990 housing units being selected. For each housing unit selected, the type of structure in which the unit was located was recorded, its damage and state of repair was assessed, and its occupancy status was determined. If the unit was occupied the household was interviewed. When necessary, multiple visits to occupied units were undertaken in order to complete interviews, although time and monetary constraints placed limits on the number of revisits. The household interview gathered information on ethnic/racial status, movement by members following the storm, insurance, and residency status of each occupant for various time periods during 1993. The interviews were conducted during the late summer and early fall of 1993, with a supplemental round of interviews conducted during December.

HOMESTEAD HOUSING NEEDS AND DEMOGRAPHIC STUDY

Tom Wilson (FIU/FAU Joint Center), Hugh Gladwin, and Walter Gillis Peacock, Co-Directors. Funded by the City of Homestead.

Population estimates produced by Florida's Bureau of Business and Economic Research (in part using data mentioned above) indicated a major drop in the population of the City of Homestead. For purposes of planning, therefore, 1990 Census data were useless, particularly when considering matters related to housing and housing quality. Faced with such a situation, the solution is to utilize sampling procedures and methods in order to obtain reliable and valid statistical information through which population estimates can be made. The City of Homestead contracted with Peacock and Gladwin (through Tom Wilson of the FIU/FAU Joint Center) to undertake a survey of housing and demographic changes. These data were collected using standard telephone survey techniques. The target population was all households located within the political boundaries of Homestead.

The sample was generated by Survey Sampling, Inc., using standard random digit dialing procedures. The nature of this technique requires the identification of telephone exchange prefixes for an area and then the generation of numbers within the prefix range. Unfortunately, the political boundaries of Homestead do not conform with telephone exchanges. As a result, likely telephone numbers were generated for all exchanges falling partially or wholly within the boundaries of Homestead. This necessitated a series of screening questions at the beginning of each interview to determine if the contacted household was indeed located within Homestead. The final sample size was approximately 950. Respondents were asked a series of questions on housing, housing markets, and housing problems. These data were then utilized by Homestead to develop housing programs, but they were also incorporated into some of the research reported here.

FLORIDA CITY STUDY

Nicole Dash, Principal Investigator.

In order to document the impact of Hurricane Andrew on Florida City, and subsequently to highlight the early stages of restoration, two spheres were analyzed – residential and commercial. For purposes of comparison, the analysis was also completed for the neighboring town of Homestead. Viewing devaluation as damage, changes in tax assessments between 1992 and 1993 were used. Vacant lots and new buildings under construction were excluded from the impact analysis. In all, 1,462 folios for Florida City and 3,724 from Homestead, both residential and commercial, were used. In addition, Dun and Bradstreet data (obtained from Conquest database) were used to assess changes in various business sectors.

In order to get a more complete picture of early restoration progress, community-level data were also collected regarding success in obtaining federal assistance, including SBA loans and Individual and Family Grants. Other data sources included interviews with officials and various reports and media stories.

EMERGENCY MANAGEMENT ORGANIZATIONAL ANALYSIS

Harvey Averch and Milan Dluhy, Principal Investigators.

From April to August 1993, Averch and Dluhy, along with Mary McDonald, conducted twenty-seven interviews with key decision makers, actors, and agents at different levels of government who had been involved in the early days of the Hurricane Andrew response. An informal conversational approach was used, but respondents were probed to solicit information about the following questions: (1) In retrospect, what were the lessons learned, i.e. what would you build into organizational planning and decision making process as a result of Andrew? (2) During the crisis period (defined as 48 hours before, during the storm, and three days after), did you or your organization deviate in any way from how you usually make decisions? (3) What kind of factors (i.e. circumstances, situations, threats, opportunities) would cause you and your organizations (also others in the network) to improvise? (4) Did a coalition of organizations emerge during the crisis? (5) During intense crises like Andrew, are there any solutions internal to the South Florida system that would make the system work better? How about solutions exogenous to South Florida? (6) Could you recommend other key intergovernmental actors at all three levels of government that we should talk with to get a better picture of decision making and management during the crisis period? All three interviewers took separate notes and then checked them against each other. Where questions arose, sources were re-contacted and, in some cases, interviewed a second time.

BIBLIOGRAPHY

Abel, E. and Nelson, M. (eds) (1990) *Circles of Care: Work and Identity in Women's Lives* (Albany, SUNY).

Abril-Ojeda, G. (1982) *The Role of Disaster Relief for Long-Term Development in LDCs* (Stockholm, University of Stockholm, Institute of Latin American Studies).

ACBJ Research Report (1995, February) Women-owned businesses. A national look, *South Florida Business Journal*, 2B.

Acker, J. (1991) Hierarchies, jobs, bodies: A theory of gendered organizations, in: J. Lorber and S. Farrell (eds) *The Social Construction of Gender*, pp. 162–179 (Newbury Park, Sage Publications, Inc.).

Ahlburg, D. A. and DeVita, C. J. (1992) *New Realities of the American Family* (Washington, DC, Population Reference Bureau, Inc.).

Alba, R. D. and Logan, J. R. (1992) Assimilation and stratification in the homeowner-ship patterns of racial and ethnic groups, *International Migration Review*, 26(4), 1314–1341.

Alexander, J. C. and Colomy, P. (1990) *Differentiation Theory and Social Change: Comparative and Historical Perspectives* (New York, Columbia University Press).

Allman, T. D. (1987) *Miami: City of the Future* (New York, Atlantic Monthly Press).

Alvarez, L. (1992a, 2 October) Tenants, landlords feud after Andrew, *Miami Herald*, 1A, 16A.

—— (1992b, 8 November) Trials of the tent people, *Miami Herald: Tropic*, 27.

—— (1992c, 25 November) Reopen tent city for homeless in S. Dade, FEMA is urged, *Miami Herald*, 1A, 6A.

Alvarez, L. and Higham, S. (1992, 16 November) The evicted: American refugees, *Miami Herald*, 1A, 6A.

Anderson, M. (1988) *Thinking About Women: Sociological Perspectives on Sex and Gender* (New York, Macmillan).

Arthur, L. (1994a, 31 July) Team effort helps families own homes, *Miami Herald: Local Section*.

—— (1994b, 22 August) Millionaire's labor of love: Helping South Dade, *Miami Herald*, 1A.

—— (1994c, 24 October) Florida City's new buildings probed, *Miami Herald*, 1B.

—— (1994d, 10 November) Florida City's building hits some snags, *Miami Herald: Neighbors*.

—— (1995, 29 March) Homestead renewal plan stirs tensions: Ethnic groups demand input, *Miami Herald: Local Section*.

Atkins, W. S., Carrasco, J., Dluhy, M. J., Gladwin, H. and Leiner, L. G. (1993) *The Elderly Population of South Dade*, technical report (Miami, Community Action Agency).

251

Averitt, R. T. (1968) *The Dual Economy* (New York, W. W. Norton).

Badain, D. (1980) Insurance redlining and the future of the urban core, *Columbia Journal of Law and Social Problems*, 16, 1–83.

Baker, E. J. (1991) Hurricane evacuation behavior, *International Journal of Mass Emergencies and Disasters*, 9(2), 287–310.

Barry, J. (1993, 28 November) Andrew's homesteaders, *Miami Herald*, 1J, 5J.

Barton, A. H. (1970) *Communities in Disaster* (Garden City, NY, Anchor, Doubleday).

Bates, F. L. (1974) Alternative models for the future of society: From the invisible to the visible hand, *Social Forces*, 53, 1–10.

—— (ed.) (1982) *Recovery, Change and Development: A Longitudinal Study of the Guatemalan Earthquake* (Athens, GA, Department of Sociology, University of Georgia).

—— (1993) The social network. Unpublished manuscript.

Bates, F. L. and Bacon, L. (1972) The community as a social system, *Social Forces*, 50, 371–379.

Bates, F. L., Farrell, T. and Glittenberg, J. K. (1979) Some changes in housing characteristics following the 1976 earthquake and their implications for future earthquake vulnerability, *International Journal of Mass Emergencies and Disasters*, 4, 121–133.

Bates, F. L., Fogleman, C., Parenton, V., Pittman, R. and Tracy, G. (1963) *The Social and Psychological Consequences of a Natural Disaster: A Longitudinal Study of Hurricane Audrey* (Washington, DC, National Research Council).

Bates, F. L. and Harvey, C. C. (1975) *The Structure of Social Systems* (New York, Gardner Press).

Bates, F. L. and Killian, C. D. (1981) The effects of the 1976 Guatemalan earthquake on earthen houses in Guatemala, in: G. W. May (ed.) *International Workshop on Earthen Buildings in Seismic Areas*, pp. 439–445 (Albuquerque, University of New Mexico Press).

Bates, F. L., Killian, C. D. and Peacock, W. G. (1984) An assessment of impact and recovery at the household level, *Journal of Ekistics*, 51(308), 439–445.

Bates, F. L. and Peacock, W. G. (1987) Disasters and social change, in: R. R. Dynes, B. DeMarchi and C. Pelanda (eds) *The Sociology of Disasters* (Milan, Italy, Franco Angeli Press).

—— (1989a) Conceptualizing social structure: The misuse of classification in structural modeling, *American Sociological Review*, 54, 565–577.

—— (1989b) Long term recovery, *International Journal of Mass Emergencies and Disasters*, 7(3), 349–365.

—— (1992) Measuring disaster impact on household living conditions: The domestic assets approach, *International Journal of Mass Emergencies and Disasters*, 10(1), 133–160.

—— (1993) *Living Conditions, Disasters, and Development: An Approach to Cross-Cultural Comparisons* (Athens, University of Georgia Press).

Bates, F. L. and Pelanda, C. (1994) An ecological approach to disasters, in: R. R. Dynes and K. J. Tierney (eds) *Disasters, Collective Behavior, and Social Organization*, pp. 149–159 (Newark, University of Delaware Press).

Bateson, G. (1972) *Steps to an Ecology of Mind* (New York, Ballantine).

Beacon Council (1994–95b) Directories: Multinationals, *Miami Business Profile*, 119–134.

BEBR (1993) *Florida Statistical Abstract, 1993*, 27th Edition (Gainesville, Bureau of Economic and Business Research, University of Florida).

—— (1994) *Florida Statistical Abstract, 1994*, 28th Edition (Gainesville, Bureau of Economic and Business Research, University of Florida).

Beck, E. M., Tolbert, C. and Horan, P. (1978) Stratification in a dual economy: A structural model of earnings determination, *American Sociological Review*, 43, 704–720.

Belcher, J. C. and Bates, F. L. (1983) Aftermath of a natural disaster: Coping through residential mobility, *Disasters*, 118–128.

Blaikie, P., Cannon, T., Davis, I. and Wisner, B. (1994) *At Risk: Natural Hazards, People's Vulnerability, and Disasters* (London, Routledge).

Bolin, R. C. (1982) *Long-Term Family Recovery from Disaster* (Boulder, University of Colorado).

—— (1984) Impact and recovery: A comparison of Black and white disaster victims (Las Cruces, Department of Sociology, New Mexico State University).

—— (1986) Disaster impact and recovery: A comparison of Black and White victims, *International Journal of Mass Emergencies and Disasters*, 4(1), 35–51.

Bolin, R. C. and Bolton, P. (1983) Recovery in Nicaragua and the USA, *International Journal of Mass Emergencies and Disasters*, 1, 124–144.

—— (1986) *Race, Religion, and Ethnicity in Disaster Recovery* (Boulder, University of Colorado).

Bolin, R. C. and Stanford, L. (1991) Shelter, housing and recovery: A comparison of US disasters, *Disasters*, 15(1), 24–34.

Bolin, R. C. and Trainer, P. (1978) Modes of family recovery following disaster: A cross-national study, in: E. L. Quarantelli (ed) *Disasters: Theory and Research* (London, Sage).

Boswell, T. D. (1992) *Ethnic Segregation in Greater Miami* (Miami, The Cuban American Policy Center, Cuban American National Council, Inc.).

—— (1993) Racial and ethnic segregation patterns in metropolitan Miami, Florida, 1980–1990, *Southeastern Geographer*, 33(1), 82–109.

Boswell, T. D. and Curtis, J. R. (1991) The Hispanization of metropolitan Miami, in: T. D. Boswell (ed.) *South Florida: The Winds of Change*, pp. 140–161 (Miami, Association of American Geographers).

Boulding, K. (1981) *Ecodynamics* (Beverly Hills, CA, Sage).

Bratt, R., Hartman, C. and Meyerson, A. (1986) *Critical Perspectives on Housing* (Philadelphia, Temple University Press).

Braudel, F. (1981) *The Structures of Every Day Life* (New York, Harper and Row).

—— (1982) *The Wheels of Commerce* (New York, Harper and Row).

—— (1984) *The Perspective of the World* (New York, Harper and Row).

Brennan, F. (1992a, 12 November) Architect urging permanent housing for farm workers, *Miami Herald*, 1A, 14A.

—— (1992b, 1 November) Florida City speaks: Residents fear, laud plan to rebuild town, *Miami Herald*, 2B.

—— (1993, 29 August) Face lift bids tenants bright welcome home, *Miami Herald: Neighbors*, 32.

Britton, N. R. (1986) Developing an understanding of disaster, *ANZJS*, 22, 254–271.

Browning, M. (1992, 27 September) Life in tent city: An indignity, a salvation, *Miami Herald*, 1A, 27A.

—— (1993, 22 August) Andrew will never be erased, *Miami Herald*, 1D, 11D.

Brydon, L. and Chant, S. (1989) *Women in the Third World: Gender Issues in Rural and Urban Areas* (New Brunswick, NJ, Rutgers).

Bureau of the Census (1990) *Survey of Minority-Owned Business Enterprises*, 1987/Blacks (Washington, US Department of Commerce).

—— (1992) *Current Population Reports* (Washington).

Cadwallader, M. (1992) *Migration and Residential Mobility: Micro and Macro Approaches* (Madison, University of Wisconsin Press).

Calvert, R. (1985) The value of biased information: A rational choice model of political advice, *Journal of Politics*, 47, 530–555.

Castanos-Lomnitz, H. (1993, August) Mexico 1985: Vulnerability and social change. Paper presented to the American Sociological Association (Miami, FL).

Catarinussi, B., Pelanda, C. and Moretti, A. (eds) (1981) *Il Disastro: Effeti Di Lungo Termine* (Udine, Grillo).

Catton, W. J. and Dunlap, R. (1978a) Environmental sociology: A new paradigm, *American Sociology*, 13, 41–49.

—— (1978b) Paradigms, theories, and the primacy of the HEP-NEP distinction, *American Sociology*, 13, 256–259.

Centers for Disease Control (1992) *Post Hurricane Andrew Assessment of Health Care Needs and Access to Health Care in Dade County, Florida.* EPI-AID 93–09 (Miami, Florida Department of Health and Rehabilitative Services).

Chowdhury, A., Mushtaque, R., Bhuyia, A. J., Choudhury, A. Y. and Sen, R. (1993) The Bangladesh cyclone of 1991: Why so many people died, *Disasters*, 17(4), 291–304.

Cigler, B. A. (1984) Emergency management and public administration, in: M. Charles and J. C. Kim (eds) *Crisis Management: A Case Book* (Springfield, IL, Charles C. Thomas).

City of Florida City (1993) *Vision 21: Florida City's Strategic Plan* (Florida City, City of Florida City).

—— (1994) *Hurricane Recovery Report: From August 24, 1992 to May 13, 1994* (Florida City, City of Florida City).

City of Homestead (1992) *Natural Disaster Plan: Annex 1, Hurricane Plan* (Homestead, City of Homestead).

Clifford, R. (1956) *The Rio Grande Flood: A Comparative Study of Border Communities*, National Research Council Disaster Study No. 7 (Washington, National Academy of Sciences).

Coble, T. A. (1995, May) Personal correspondence to Betty Morrow (Miami).

Cochrane, H. C. (1975) *Natural Hazards and Their Distributive Effect* (Boulder, University of Colorado).

Colina, D. G. (1995, 24 August) Personal communication (Miami, FL).

Collins Center for Public Policy (1995) *Final Report: Academic Task Force on Hurricane Catastrophe Insurance* (Tallahassee, FL, Collins Center for Public Policy).

Collins, P. H. (1990) *Black Feminist Thought: Knowledge, Consciousness, and the Politics of Empowerment* (New York, Routledge).

Cronin, F. (1982) Racial differences in the search for housing, in: W. Clark (ed.) *Modeling Housing Market Search*, pp. 81–105 (London, Croom Helm).

Dacy, D. C. and Kunreuther, H. (1969) *The Economics of Natural Disasters* (New York, Free Press).

Dade County Commission on the Status of Women (1995) *Local Women Government Officials* (Miami, Metro Dade).

Dahl, R. A. (1961) *Who Governs?* (New Haven, Yale University Press).

deBarbieri, M. T. and Guzman, C. (1986) Las damnificadas y el empleo, *Revista Mexicana de Sociologia*, 48(2), 59–101.

Didion, J. (1987) *Miami* (New York, Simon and Schuster).

Dluhy, M. J. (1990) *Building Coalitions in the Human Services* (Newbury Park, CA, Sage).

—— (1995, 13 July) Dade County's incorporation fever, *Miami Herald: Neighbors.*

Domestic Violence Coordination Unit (1995) 1995 Report of Domestic/Repeat Violence Injunctions (Miami, 11th Judicial Circuit of Florida).

Donnelly, J. (1993, 22 August) We Will Rebuild critics concede benefits of clout, *Miami Herald; Special Report: Out of the Storm.*

Drabek, T. E. (1969) Social processes in disaster: Family evacuation, *Social Problems*, 16, 336–349.

—— (1983) Shall we leave? A study on family reactions when disaster strikes, *Emergency Management Review*, 1, 25–29.

—— (1986) *Human Responses to Disaster* (New York, Springer-Verlag).

Drabek, T. E. and Boggs, K. S. (1968, August) Families in disaster: Reactions and relatives, *Journal of Marriage and the Family*, 443–451.

Drabek, T. E. and Key, W. H. (1976) The impact of disaster on primary group linkages, *Mass Emergencies*, 1, 89–105.

—— (1982) *Conquering Disaster: Family Recovery and Long-Term Consequences* (New York, Irvington).

Drabek, T. E., Key, W. H., Erickson, P. E. and Crowe, J. L. (1975) The impact of disaster on kin relationships, *Journal of Marriage and the Family*, 481–494.

Due, T. (1993a, 7 March) The people who fall through the cracks, *Miami Herald*, 1J, 4J.

—— (1993b, 9 August) Pregnancy and the strain of Andrew, *Miami Herald*, 1C, 2C.

Dunal, C., Gaviria, M., Flaherty, J. and Birz, S. (1985) Perceived disruption and psychological stress among flood victims, *Journal of Operational Psychiatry*, 16(2), 9–16.

Duncan, O. (1961) From social system to ecosystem, *Sociological Inquiry*, 31, 140–146.

—— (1964) Social organization and the ecosystem, in: R. Faris (ed.) *Handbook of Modern Sociology*, pp. 36–82 (Chicago, Rand McNally).

Dunlap, M. (1992, 3 November) Designers unveil plan for a new Florida City, *Miami Herald*.

Dunlap, R. E. and Catton, W. J. (1979) Environmental sociology, *Annual Review of Sociology*, 5, 243–273.

Dunn, M. and Stepick, A., III (1992) Blacks in Miami, in: G. J. Grenier and A. Stepick, III (eds) *Miami Now! Immigration, Ethnicity, and Social Change*, pp. 41–56 (Gainesville, University Press of Florida).

Dynes, R. R. (1974) *Organized Behavior in Disasters* (Lexington, MA, D.C. Heath and Co.).

—— (1975) The comparative study of disaster: A social organizational approach, *Mass Emergencies*, 1, 21–31.

Epstein, C. F. (1988) *Deceptive Distinctions: Sex, Gender, and the Social Order* (New Haven, Yale University Press).

Erickson, P. E., Drabek, T. E., Key, W. H. and Crowe, J. L. (1976) Families in disaster: Patterns of recovery, *Mass Emergencies*, 1, 203–216.

Erikson, K. T. (1976) *Everything in Its Path: Destruction of Community in the Buffalo Creek Flood* (New York, Simon and Schuster).

—— (1994) *A New Species of Trouble: Explorations in Disaster, Trauma, and Community* (New York, W. W. Norton and Company).

Fardon, R. (ed.) (1990) *Socializing Strategies: Regional Traditions of Ethnographic Writing* (Washington, Smithsonian Press).

Faupel, C. E. (1985) *The Ecology of Disaster: An Application of a Conceptual Model* (New York, Irvington).

—— (1987) Human ecology and disaster: Contributions to research and policy formation, in: R. R. Dynes, B. DeMarchi and C. Pelanda (eds) *Sociology of Disasters: Contribution of Sociology to Disaster Research*, pp. 181–210 (Milan, Franco Angeli).

FEMA (1990) *Response to Hurricane Hugo and the Loma Prieta Earthquake: Evaluation and Lessons Learned* (Washington, Federal Emergency Management Agency).

—— (1992a) *Federal Response Plan* (Washington, Federal Emergency Management Agency).

—— (1992b) *Hurricane Andrew Situation Reports 1–15* (Washington, Federal Emergency Management Agency).

—— (1993a) *Andrew, Iniki, Omar: Evaluation of Federal Response and Recovery Efforts* (Washington, Federal Emergency Management Agency).

—— (1993b) *FEMA's Disaster Management Program: A Performance Audit After Hurricane Andrew*, Inspector General report (Washington, Federal Emergency Management Agency).

—— (1994, May 5) *Mobile home operations daily report*, technical report (Miami, Federal Emergency Management Agency, Disaster Field Office 955).

Filkins, D. (1993, 8 September) Plan to aid S. Dade endorsed, *Miami Herald*, 1B, 2B.

—— (1994, 16 December) A new leader for Metro: Latin bloc prevails in manager vote, *Miami Herald*, 1A.

Finch, J. and Groves, D. (1983) *A Labour of Love: Women, Work, and Caring* (London, Routledge and Kegan Paul).

Fitzpatrick, C. and Mileti, D. S. (1991) Motivating public evacuation, *International Journal of Mass Emergencies and Disasters*, 9(2), 137–152.

Fjellman, S. M. (1992) *Vinyl Leaves: Walt Disney World and America* (Boulder, CO, Westview).

Fogelman, C. and Parenton, V. (1956) Disaster and aftermath: Selected aspects of individual and group behavior in critical situations, *Social Forces*, 38, 130.

Forest, T. R. (1978) Group emergence in disasters, in: E. Quarantelli (ed.) *Disasters: Theory and Research*, pp. 105–129 (Beverly Hills, CA, Sage).

Fothergill, A. (1996) Gender, risk and disaster, *International Journal of Mass Emergencies and Disasters*, 14(1), 33–55.

Fried, M. (1966) Grieving for a lost home, in: J. Wilson (ed) *Urban Renewal: The Record and the Controversy* (Cambridge, MA, MIT Press).

Friesema, H. P., Caporaso, J., Goldstein, G., Lineberry, R. and McCleary, R. (1979) *Aftermath: Communities After Natural Disasters* (Beverly Hills, CA, Sage).

Fronstin, P. and Holtmann, A. G. (1994) The determinants of residential property damage caused by Hurricane Andrew, *Southern Economic Journal*, 61(2), 387–397.

GAO (1993) *Disaster Management: Recent Disasters Demonstrate the Need to Improve the Nation's Response Strategy* (Washington, General Accounting Office).

Geipel, R. (1982) *Disaster and Reconstruction* (London, George Allen and Unwin).

Getter, L. (1992a, 20 December) Inspections: A breakdown in the system, *Miami Herald*, 6SR–7SR.

—— (1992b, 20 December) Do builders' bucks buy political power? *Miami Herald*, 7SR.

Gillespie, D. F. and Colignon, R. A. (1993) Structural change in disaster preparedness networks, *International Journal of Mass Emergencies and Disasters*, 11(2), 143–162.

Gillespie, D. F., Colignon, R. A., Banerjee, M. M., Murty, S. A. and Rogge, M. (1992) *Interorganizational Relations for Disaster Preparedness*, Final report (St. Louis, MO, Washington University).

Gladwin, H. (1993) *Emergency Assistance and Relief Activities: Immediate Aftermath*, Proceedings from Lessons Learned from Hurricane Andrew Conference (Miami, Florida International University).

Gluckman, M. (1958 [1940–1942]) *Analysis of a social situation in modern Zululand*, Rhodes-Livingston Papers, No. 28 (Manchester, Manchester University Press for the Rhodes-Livingston Institute).

Goldscheider, F. K. and Waite, L. J. (1991) *New Families, No Families? The Transformation of the American Home* (Berkeley, University of California Press).

Gore, R. (1993, April) Andrew aftermath, *National Geographic*, 183(4), 2–37.

Governor's Disaster Planning and Response Review Committee (1993) *Final Report*, (Tallahassee, State of Florida).

Grenier, G. J. and Stepick, A., III (eds) (1992) *Miami Now! Immigration, Ethnicity, and Social Change* (Gainesville, University Press of Florida).

Guillette, E. A. (1993) *The Role of the Aged in Community Recovery Following Hurricane Andrew*. QR56. Quick Response Reports (Boulder, CO, Natural Hazards Research and Applications Information Center).

Guy, R. F., Pol, L. G. and Ryker, R. (1982) Discrimination in mortgage lending: The Mortgage Disclosure Act, *Population Research and Policy Review*, 1, 283–296.

Haas, J. E., Kates, R. W. and Bowden, M. J. (1977) *Reconstruction Following Disaster* (Cambridge, MA, The MIT Press).

Hale, K. C. (1993, 15 September) Andrew: One year later. Institute for Public Policy and Citizenship Studies, Faculty Lunchtime Symposium (Miami: Florida International University).

Hancock, D. (1993, 28 May) Lawsuit: FEMA didn't provide adequate housing after storm, *Miami Herald*, 3B.

Hannan, M. T. and Freeman, J. (1989) *Organizational Ecology* (Cambridge, MA, Harvard University Press).

Harmon, H. H. (1967) *Modern Factor Analysis* (Chicago, University of Chicago Press).

Hartman, T. (1992, 24 October) S. Dade's last tent city folds up, *Miami Herald*, 1A, 18A.

—— (1993a, 2 April) S. Dade rental housing still lacking, *Miami Herald: Neighbors*.

—— (1993b, 9 May) FEMA calls for patrols at trailers, *Miami Herald: Neighbors*, 30.

—— (1993c, 16 February) Rough road back: Florida City is climbing out of chaos of storm's aftermath, *Miami Herald*, 1B, 4B.

—— (1994, 14 August) Citizens seek a new destiny, *Miami Herald: Neighbors*.

Hartman, T. and Penn, I. (1992, 6 December) Apartment shortage hits South Florida, *Miami Herald*, 1B, 3B.

Hebert, P. J., Jarrell, J. D. and Mayfield, M. (1993) *The Deadliest, Costliest and Most Intense Hurricane of This Century (and Other Frequently Requested Hurricane Facts)*, NOAA Technical Memorandum NWS NNHC31 (Coral Gables, FL, NOAA).

—— (1996) *The Deadliest, Costliest, and Most Intense United States Hurricanes of This Century (and Other Frequently Requested Hurricane Facts)*. NOAA Technical Memorandum NWS TPC-1 (Miami, Florida International University, Tropical Prediction Center).

Herman, M. (1995, August) A tale of two cities: Testing explanations for riot violence in Miami Florida and Los Angeles California 1980, 1992, American Sociological Association (Washington, DC).

Hess, B. and Ferree, M. M. (1987) *Analyzing Gender: A Handbook of Social Science Research* (Beverly Hills, CA, Sage).

Hewitt, K. (1983) *Interpretations of Calamity: From the Viewpoint of Human Ecology* (London, Allen and Unwin, Inc.).

Higham, S. (1992a, 9 September) Victims lose trailers, faith in Feds, *Miami Herald*, 1A, 21A.

—— (1992b, 16 October) Last of U.S. military pulls out of storm area, *Miami Herald*, 1B.

Hill, R. and Hansen, D. A. (1962) Families in disaster, in: G. Baker and D. Chapman (eds) *Man and Society in Disaster*, pp. 185–221 (New York, Basic Books).

Hill, R. C. (1974) Separate and unequal: Governmental inequality in the metropolis, *American Political Science Review*, 68, 1557–1568.

Historic Publications (1992) *Hurricane Andrew Documentary* (Charleston, SC, Historic Publications).

Hochschild, A. (1989) *The Second Shift: Working Parents and the Revolution at Home* (New York, Viking).

257

Hoffman, S. M. (1993) Up from the embers: The aftermath of the Oakland Berkeley firestorm: A survivor anthropologist's perspective, Paper presented at the Society for Applied Anthropology annual meeting (San Antonio, TX).

Hoover, G. A. and Bates, F. L. (1985) The impact of a natural disaster on the division of labor in twelve Guatemalan communities: A study of social change in a developing country, *International Journal of Mass Emergencies and Disasters*, 3(3), 9–26.

Horton, H. D. (1992) Race and wealth: A demographic analysis of Black home-ownership, *Sociological Inquiry*, 62(4), 480–489.

Hunter, F. (1953) *Community Power Structure: A Study of Decision Makers* (Chapel Hill, University of North Carolina Press).

Immigration and Naturalization Service (1948) *Annual Report* (Washington, US Department of Justice).

Institute of Government (1993) *Improving Decision Making in the Aftermath of Hurricane Andrew* (Miami: Florida International University).

Jones, R. T., Frary, R., Cunningham, P. and Weddle, J. D. (1993) *The Psychological Effects of Hurricane Andrew on Elementary and Middle School Children* (Boulder, CO, Natural Hazards Research and Applications Information Center).

Kaniasty, K. and Norris, F. H. (1993) A test of social support deterioration model in the context of natural disaster, *Journal of Personality and Social Psychology*, 64(3), 395–408.

Keen, R. A. (1995) *Dissolutions of Marriage Filed in Dade County, Florida: January 1990 – February 1995* (Miami, Family Division, Circuit and County Courts).

Kerr, O. (1984) *Profile of the Black Population* (Miami, Metro Dade Planning Department).

Khondker, H. H. (1996) Women and floods in Bangladesh, *International Journal of Mass Emergencies and Disasters*, 14(3), 281–92.

Killian, C. D., Peacock, W. G. and Bates, F. L. (1984) The inequality of disasters: An assessment of the interaction between a social system and its geophysical environment, Southern Sociological Society (Knoxville, TN).

Kish, L. (1965) *Survey Sampling* (New York, John Wiley).

Klinteberg, R. (1979) Management of disaster victims and rehabilitation of uprooted communities, *Disasters*, 3(1), 61–70.

Kreps, G. A. (1984) Sociological inquiry and disaster research, *Annual Review of Sociology*, 10, 309–330.

——— (1989) *Social Structure and Disaster* (Newark, University of Delaware and Associated University Presses).

LaGreca, A. M., Silverman, W. K., Vernberg, E. and Prinstein, M. J. (1996) Symptoms of posttraumatic stress symptoms in children following Hurricane Andrew: A prospective study, *Journal of Consulting and Clinical Psychology*, 105, 712–23.

Lake, R. W. (1980) Racial transition and Black homeownership in American suburbs, in: G. Sternlieb and J. W. Hughes (eds) *America's Housing*, pp. 419–438 (New Brunswick, NJ, Center for Urban Policy Research).

Laska, S. (1993) Environmental sociology and the state of the discipline, *Social Forces*, 72(1), 1–17.

Laudisio, G. (1993) Disaster aftermath: Redefining response – Hurricane Andrew's impact on I & R, *Alliance of Information and Referral Systems*, 15, 13–32.

League of Red Cross and Red Crescent Societies (1991) *Working with Women in Emergency Relief and Rehabilitation Programmes*, Field Studies Paper No. 2 (Geneva, Switzerland: League of Red Cross).

Leahy, P. (1985) Are racial factors important for the allocation of mortgage money? A quasi-experimental approach to an aspect of discrimination, *American Journal of Economics and Sociology*, 44, 185–197.

258

Leavitt, J. (1992) Women under fire: Public housing activism in Los Angeles, *Frontiers* 13(2), 109–30.

Leen, J., Doig, S. K. and Getter, L. (1993, 20 December) Failure of design and discipline, *Miami Herald: Special Report*, 2–3.

Lenski, G. and Lenski, J. (1990) *Human Societies* (New York, McGraw Hill).

Logan, J. R. and Molotch, H. L. (1987) *Urban Fortunes: The Political Economy of Place* (Berkeley, University of California Press).

Loudner, R. (1992, October) Andrew: A mother's story, *South Florida Parenting*, 10–12.

Luhmann, N. (1990a) *Essays on Self Reference* (New York, Columbia University Press).

—— (1990b) The paradox of system differentiation and the evolution of society, in: J. C. Alexander and P. Colomy (eds) *Differentiation Theory and Social Change: Comparative and Historical Perspectives*, pp. 409–440 (New York, Columbia University Press).

Marks, M. (1993, 15 June) Storm displaced students homesick for Homestead, *Miami Herald*, 1A, 15A.

Massey, D. S. and Denton, N. A. (1993) *American Apartheid* (Cambridge, MA, Harvard University Press).

Massolo, A. and Schteingart, M. (1987) *Participacion social, reconstruccion y mujer. El sismo de 1985* (Mexico City, El colegio de Mexico).

Maturana, H. and Varela, F. (1980) *Autopoiesis and Cognition: The Realization of the Living* (Boston, Reidel).

Merton, R. (1957) *Social Structure and Social Theory* (New York, The Free Press).

Merzer, M. and Viglucci, A. (1992, 22 November) Advocates for homeless: 'What we need is action,' *Miami Herald*, 1A, 28A.

Metro Dade (1992) *Hurricane Procedure* (Miami, Metro Dade County).

—— (1993) *A Minority– and Women-Owned Business Discrimination Study*, Executive summary (Miami, Metro Dade County).

—— (1994a) *Black Elected Officials of Dade County* (Miami, Metro Dade County, Office of Black Affairs).

—— (1994b) *Dade County Project Chart Reports: Schedule of Repair, Completion of Uninhabitable Units* (Miami, Metro Dade County).

Metro Dade Board of County Commissioners (1990) *Metropolitan Dade County at-a-Glance* (Miami, Metro Dade Communications Department).

Metro Dade Housing and Urban Development (1994, April 14) *Report: completion schedule of uninhabitable units* (Miami, Metro Dade County).

Metro Dade Office of Emergency Management (1992) *Metropolitan Dade County Emergency Operations Plan* (Miami, Metro Dade County).

Metro Dade Planning Department (1992) *Hurricane Andrew: Impact Area Profile*, technical report (Miami, Metro Dade County).

—— (1993) *Population Estimates and Projections: Post-Hurricane Andrew, Dade County, Florida*, report (Miami, Metro Dade County).

—— (1994) *Housing and Population Recovery Post-Hurricane Andrew* (Miami, Metro Dade County).

Miami Herald (1992a) *The Big One: Hurricane Andrew* (Miami, *Miami Herald* Publishing Company).

—— (1992b, 28 August) Where the hell is the cavalry on this one? *Miami Herald*, 1A.

—— (1992c, 4 September) Tents up, but few eager to move in, *Miami Herald*, 19A.

—— (1992d, 5 September) Tent city seen as last resort, *Miami Herald*, 1A, 22A.

—— (1992e, 5 September) As tent cities open: "It's gonna be rough," *Miami Herald*, 1A.

—— (1992f, 9 September) Officers pulled from taking food to migrants, *Miami Herald*, 23A.

—— (1992g, 12 September) Some refuse to evacuate trailer parks, *Miami Herald*, 23A.

—— (1992h, 12 September) Number of relief workers plunges, *Miami Herald*, 1A, 24A.

—— (1992i, 27 September) Andrew – Four weeks and counting, *Miami Herald*, 1A, 22A.

—— (1992j, 29 September) Farmworker relief center comes to life, *Miami Herald*, 1A, 17A.

—— (1992k, 7 October) Storm weary parents give up adopted kids, *Miami Herald*, 1A, 10A.

—— (1992l, 8 November) Planned Parenthood provides birth control to storm victims, *Miami Herald: Neighbors*, SS 2.

—— (1992m, 25 November) Migrant workers arriving, *Miami Herald*, 6A.

—— (1992n, 20 December) Special report: What went wrong? *Miami Herald*, SR 1–16.

—— (1993a, 10 May) FEMA taking applications again for storm aid, *Miami Herald*, 4B.

—— (1993b, 22 August) Andrew by the numbers, *Miami Herald*, 10D.

—— (1993c, 7 November) Rentals still hard to find in some areas, *Miami Herald*, 1B, 4B.

—— (1994, 24 August) Andrew recovery by the numbers, *Miami Herald*, 18A.

—— (1995, 23 August) The emergency within, *Miami Herald*, 18A.

Mileti, D. S. (1975) *Natural Hazard Warning Systems in the United States: A Research Assessment* (Boulder, Institute of Behavioral Science, The University of Colorado).

Mileti, D. S., Drabek, T. E. and Haas, J. E. (1975) *Human Systems in Extreme Environments: A Sociological Perspective* (Boulder, University of Colorado, Institute of Behavioral Science, Program on Environment and Behavior, Monograph No. 21).

Mileti, D. S., Haas, J. E. and Gillespie, D. F. (1977) Size and structure in complex organizations, *Social Forces*, 56, 208–217.

Mills, C. W. (1956) *The Power Elite* (New York, Oxford University Press).

Mitrani, J. (1993) Introduction to Construction and Design Session, in: P. H. Mann (ed.) *Lessons Learned from Hurricane Andrew: Conference Proceedings*, pp. 24–27 (Miami, Florida International University).

Mohl, R. (1983) Miami: The ethnic cauldron, in: R. Bernard and B. Rice (eds) *Sunbelt Cities*, pp. 67–72 (Austin, University of Texas Press).

—— (1985) An ethnic 'boiling pot': Cubans and Haitians in Miami, *Journal of Ethnic Studies*, 13(2), 51–74.

—— (1991) The settlement of Blacks in South Florida, in: T. D. Boswell (ed.) *South Florida: The Winds of Change*, pp. 112–139 (Miami, Association of American Geographers).

Moore, H. E. (1958) *Tornadoes Over Texas* (Austin, University of Texas Press).

Moore, H. E., Bates, F. L. and Parenton, V. (1963) *Before the Wind* (Washington, National Academy of Sciences).

Morrow, B. H. (1992, May) The aftermath of Hugo: Social effects on St. Croix, Caribbean Studies Association (St. George's, Grenada).

—— (1993, March) Families of Andrew, 1993 Meeting of the Society for Applied Anthropology (San Antonio, TX).

Morrow, B. H. and Enarson, E. (1994, July) Making the case for gendered disaster research, XIIIth World Congress of Sociology (Bielefeld, Germany).

—— (1996) Hurricane Andrew through women's eyes: Issues and recommendations, *International Journal of Mass Emergencies and Disasters*, 14(1), 1–22.

Morrow, B. H. and Peacock, W. G. (1993) The social impact of Hurricane Andrew, in: P. H. Mann (ed.) *Lessons Learned from Hurricane Andrew: Conference Proceedings* (Miami, Florida International University).

Morrow, B. H., Peacock, W. G. and Enarson, E. (1994) *Assessing a Community Recovery Function for the ARC Disaster Response Plan* (Alexandria, VA, American Red Cross Hurricane Andrew Recovery Project).

Morrow, B. H. and Ragsdale, A. K. (1996) *Early Response to Hurricane Marilyn in the U.S. Virgin Islands*, Quick Response Research Report No. 82 (Boulder, CO, Natural Hazards Center, University of Colorado).

Morrow-Jones, H. A. and Morrow-Jones, C. R. (1991) Mobility due to natural disaster: Theoretical considerations and preliminary analysis, *Disasters*, 15(2), 126–132.

Musibay, O. (1995, 8 July) Funding to help rebuild 1,000 storm-hit homes, *Miami Herald*, 1B, 3B.

NAPA (1993) *Coping with Catastrophe: Building an Emergency Management System to Meet People's Needs in Natural and Manmade Disasters* (Washington, National Academy of Public Administration).

Neal, D. M. (1984) Blame assignment in a diffuse disaster situation: A case example of the role of an emergent citizen group, *International Journal of Mass Emergencies and Disasters*, 2(2), 251–266.

Neal, D. M. and Phillips, B. (1990) Female-dominated local social movement organizations in disaster-threat situations, in: G. West and R. L. Blumberg (eds) *Women and Social Protest*, pp. 243–255 (New York, Oxford University Press).

Neal, D., Perry, J., Joseph, B. and Hawkins, R. (1982) Getting ready for blizzards: Preparation levels in the winter of 1977–78, *Sociological Focus*, 15(1), 67–76.

New South Dade Planning Charrette (1993) *The New South Dade Planning Charrette: From Adversity to Opportunity*, Planning Report (Miami, Florida International University, School of Design and The University of Miami).

Newton, K. (1976) Feeble governments and private power: Urban politics and policies in the United States, in: L. Masottie and R. Lineberry (eds) *The New Urban Politics*, pp. 37–56 (Cambridge, MS, Ballinger).

Nigg, J. M. and Perry, R. W. (1988) Influential first sources: Brief statements with long-term effects, *International Journal of Mass Emergencies and Disasters*, 6(3), 311–343.

Nigg, J. and Tierney, K. (1990) *Explaining Differential Outcomes in the Small Business Disaster Loan Application Process*, Preliminary paper no. 156 (Newark, University of Delaware, Disaster Research Center).

Nix, H. L., Dressel, P. L. and Bates, F. L. (1977) Changing leaders and leadership structure: A longitudinal study, *Rural Sociology*, 42(1), 22–41.

Odum, H. (1971) *Environment, Power, and Science* (New York, Wiley-Interscience).

Office of County Manager (1994) *Homestead Air Force Base and South Dade Business Development and Marketing Plan, November 1, 1994* (Miami, Metro Dade County).

Office of Vital Statistics (1995) *Public Health Statistics* (Jacksonville, Florida Department of Health and Rehabilitative Services).

O'Hare, W. P. (1987, November) Best metros for Hispanic businesses, *American Demographics*, 31–33.

—— (1992) *America's Minorities – The Demographics of Diversity*, pamphlet (Washington, Population Reference Bureau, Inc.).

Oliver-Smith, A. (1986) *The Martyred City: Death and Rebirth in the Andes* (Albuquerque, University of New Mexico Press).

—— (1990) Post-disaster housing reconstruction and social inequality: A challenge to policy and practice, *Disasters*, 14(1), 7–19.

Park, R. (1936) Human ecology, *American Journal of Sociology*, 42, 1–15.

Parker, J. F., Bahrick, L., Lundy, B., Fivush, R. and Levitt, M. (1997) Effects of stress on children's memory for a natural disaster, in: C. P. Thompson, D. J. Herrmann, J. D. Read, D. Bruce, D. G. Payne and M. P. Toglia (eds) *Autobiographical and Eyewitness Memory: Theoretical and Applied Perspectives* (Hillsdale, NJ, Erlbaum).

261

Parks, A. M. (1991) *Miami: The Magic City* (Miami, Centennial Press).

Peacock, W. G. (1991) In search of social structure, *Sociological Inquiry*, 61(3), 281–298.

—— (1994) War and accountability in Guatemala, *Hemisphere*, 6(1), 44–46.

—— (1996) Disasters, development, and mitigation: Taking a proactive stance, *Natural Hazards Observer*, XX(4), 1–2.

Peacock, W. G. and Bates, F. L. (1982) Ethnic differences in earthquake impact and recovery, in: F. L. Bates (ed.) *Recovery, Change and Development: A Longitudinal Study of the Guatemalan Earthquake*, pp. 792–892 (Athens, University of Georgia Department of Sociology).

Peacock, W. G. and Gladwin, H. (1995) *Racial and Ethnic Comparisons of Housing Characteristics, Perceptions of Rental and Housing Markets, and Problems Obtaining Housing*, Preliminary report No. 3: Homestead Housing Needs and Demographic Data Project (Miami: FIU/FAU Joint Center for Urban and Regional Planning).

Peacock, W. G., Gladwin, H. and Girard, C. (1993) *Ethnic and racial findings from the FIU Hurricane Andrew survey*, Preliminary report No. 6, Technical report (Miami, Florida International University).

Peacock, W. G., Killian, C. D. and Bates, F. L. (1987) The effects of disaster damage and housing aid on household recovery following the 1976 Guatemalan earthquake, *International Journal of Mass Emergencies and Disasters*, 5, 63–88.

Peacock, W. G., Killian, C. D., Hoover, G. A. and Bates, F. L. (1984) Alterations in community complexity and household recovery following the 1976 Guatemalan earthquake, Annual meeting of the American Sociological Association (San Antonio, TX).

Pearce, D. (1979, February) Gatekeepers and homeseekers: Institutional patterns in racial steering, *Social Problems*, 26(3), 325–342.

Pedraza-Bailey, S. (1985) *Political and Economic Migrants in America: Cubans and Mexicans* (Austin, University of Texas Press).

Pelanda, C. (1982, August) Disaster and order: Theoretical problems in disaster research, Tenth World Congress of Sociology (Mexico City, Mexico).

—— (1989) *Cognitive Neo-Systemics: Theory of Artificial Observers* (Athens, INTERLAB, University of Georgia).

Pérez, L. (1985) The Cuban population of the United States: the results of the 1980 U.S. census of population, *Cuban Studies/Estudios Cubanos*, 15, 1–18.

—— (1992) Cuban Miami, in: G. J. Grenier and A. Stepick, III (eds) *Miami Now! Immigration, Ethnicity, and Social Change*, pp. 83–108 (Gainesville, University Press of Florida).

—— (1992, 27 September) Hurricane has severely tilted community demographics, *Miami Herald*, 4M.

Perry, R. W. (1979) Evacuation decision-making in natural disasters, *Mass Emergencies*, 4, 25–38.

—— (1985) *Comprehensive Emergency Management: Evacuating Threatened Populations* (Greenwich, CT, JAI Press, Inc.).

—— (1987) Disaster preparedness and response among minority citizens, in: R. R. Dynes, B. DeMarchi and C. Pelanda (eds) *Sociology of Disasters: Contribution of Sociology to Disaster Research* (Milan, Italy, Franco Angeli Press).

Perry, R. W. and Greene, M. R. (1983) *Citizen Response to Volcanic Eruptions: The Case of Mount St. Helens* (New York, Irvington Publishers).

Perry, R. W. and Lindell, M. K. (1991) The effects of ethnicity on evacuation decision-making, *International Journal of Mass Emergencies and Disasters*, 9(1), 47–68.

Perry, R. W., Lindell, M. K. and Greene, M. R. (1980) *The Implications of Natural Hazard Evacuation Warning Studies for Crisis Relocation Planning* (Seattle, Batelle Human Affairs Research Centers).

—— (1981) *Evacuation Planning in Emergency Management* (Lexington, MA, Lexington Books).

Perry, R. W. and Mushkatel, A. H. (1986) *Minority Citizens in Disasters* (Athens, University of Georgia Press).

Peters, T. (1984) *Miami 1909* (Miami, Banyan Books).

Peterson, S. and Runyan, A. S. (1993) *Global Gender Issues* (Boulder, CO, Westview).

Phifer, J. (1990) Psychological distress and somatic symptoms after natural disaster: Differential vulnerability among older adults, *Psychology and Aging*, 5(3), 412–420.

Phillips, B. D. (1993a, August) Creating, sustaining, and losing place: Homelessness and the sociology of places, American Sociological Association (Miami, Florida).

—— (1993b) Cultural diversity in disasters: Sheltering, housing, and long-term recovery, *International Journal of Mass Emergencies and Disasters*, 11(1), 99–110.

Phillips, B. D., Garza, L. and Neal, D. M. (1994) Intergroup relations in disasters: Service delivery barriers after Hurricane Andrew, *The Journal of Intergroup Relations*, XXI(3), 18–27.

Polny, M. J. (1993) *Recovery After Action Report*, FEMA-0955–DR-FL (Atlanta, GA, Federal Emergency Management Agency).

Poniatowska, E. (1995) *Nothing, Nobody: The Voices of the Mexico City Earthquake* (Philadelphia, PA, Temple University Press).

Porter, B. and Dunn, M. (1984) *The Miami Riot of 1980: Crossing the Bounds* (Lexington, MA, D. C. Heath).

Portes, A. (1981) Modes of structural incorporation and present theories of labor immigration, in: M. M. Kritz, C. B. Keely and S. M. Tomasi (eds) *Global Trends in Migration: Theory and Research of International Population Movements*, pp. 279–297 (New York, Center for Migration Studies).

—— (1987) The social origins of the Cuban enclave economy of Miami, *Sociological Perspectives*, 30, 340–371.

—— (1992) Foreward, in: G. J. Grenier and A. Stepick, III (eds) *Miami Now! Immigration, Ethnicity and Social Change*, pp. xiii–xvi (Gainesville, University Press of Florida).

Portes, A. and Bach, R. L. (1985) *Latin Journey: Cuban and Mexican Immigrants in the United States* (Berkeley, University of California Press).

Portes, A. and Rumbaut, R. G. (1990) *Immigrant America: A Portrait* (Berkeley and Los Angeles, University of California Press).

Portes, A. and Sensenbrenner, J. (1993) Embeddedness and immigration: Notes on the social determinants of economic action, *American Journal of Sociology*, 98, 1320–1350.

Portes, A. and Stepick, A., III (1993) *City on the Edge: The Transformation of Miami* (Berkeley, University of California Press).

Prince, S. H. (1920) *Catastrophe and Social Change* (New York, Longmans, Green and Co.).

Provenzo, E. F., Jr. and Fradd, S. H. (1995) *Hurricane Andrew, the Public Schools, and the Rebuilding of Community* (Albany, State University of New York Press).

Quarantelli, E. L. (1960, August) A note on the protective function of the family in disasters, *Journal of Marriage and Family Living*, 263–264.

—— (1980) *Evacuation Behavior and Problems: Findings and Implications from the Research Literature* (Columbus, Disaster Research Center, The Ohio State University).

—— (1987) What should we study? Questions and suggestions for researchers about the concept of disasters, *International Journal of Mass Emergencies and Disasters*, 5(1), 7–32.

Rabell, C. and Teran, M. (1986) "Los daminificados por los sismos de 1985 en la Ciudad de Mexico, *Revista Mexicana de Sociologia*, 48(2), 3–28.

Reskin, B. and Padavic, I. (1994) *Women and Men at Work* (Thousand Oaks, CA, Pine Forge Press).

Reveron, D. (1989, 13 February) Violence, delays hurt renewal in Black Dade, *Miami Herald*, 1A, 6A.

Rieff, D. (1987) *Going to Miami: Exiles, Tourists and Refugees in the New America* (Boston, Little, Brown).

Rogers, P. (1993, 9 August) House calls give Andrew's victims a lift, *Miami Herald*, 1B, 2B.

Rosenthal, U. and t'Hart, P. (1991) Experts and decision makers in crisis situations, *Knowledge*, 12, 350–372.

Rossi, P. H., Wright, J. D. and Wright, S. R. (1981) Assessment of research on natural hazards reassessed in light of the SADRI Disaster Research Program, in: J. D. Wright and P. H. Rossi (eds) *Social Science and Natural Hazards*, pp. 143–159 (Cambridge, MA, Abt).

Rotton, J., Dubitsky, S. S., Milov, A. and White, S. M. (In press) Distress, elevated cortisol, cognitive deficits, and illness following a natural disaster, *Journal of Environmental Psychology*.

Rubin, C. B. (1985) The community recovery process in the United States after a major disaster, *International Journal of Mass Emergencies and Disasters*, 3, 9–28.

Sagalyn, L. B. (1983) Mortgage lending in older urban neighborhoods: Lessons from past experience, *Annals of American Academy*, 465, 98–108.

Scanlon, J. (1988) Winners and losers: Some thoughts about the political economy of disasters, *International Journal of Mass Emergencies and Disasters*, 6(1), 47–63.

Schneider, R. (1993, October) Emergency management systems coordinators as political actors, 1993 Southeastern Conference for Public Administration (Cocoa Beach, FL).

Schneider, S. K. (1992) Governmental response to disasters: The conflict between bureaucratic procedures and emergent norms, *Public Administration Review*, 52, 135–143.

Schroeder, R. (1987) *Gender Vulnerability to Drought: A Case Study of the Hausa Social Environment* (Boulder, University of Colorado, Institute of Behavioral Science).

Scott, H. (1984) *Working Your Way to the Bottom: The Feminization of Poverty* (London, Pandora Press).

Semple, K. (1993, 19–23 June) A tale of tent city, *New Times* (Miami, FL), 20, 23–24, 26, 28–30, 32, 34.

Senior, D. (1970) *Skopje Resurgent: The Story of a United Nations Special Fund Town Planning Project*, UN Document DP/SF/UN17 (New York, United Nations).

Shaw, R. (1989) Living with floods in Bangladesh, *Anthropology Today*, 5(1), 11–13.

Sheets, R. C. (1990) The National Hurricane Center – Past, present, and future, *Weather and Forecasting*, 5, 185–232.

—— (1993a) Opening Address, in: P. H. Mann (ed.) *Lessons Learned from Hurricane Andrew: Conference Proceedings*, pp. 5–15 (Miami, Florida International University).

—— (1993b, September) Personal communication.

Shelton, B. A. (1992) *Women, Men, and Time: Gender Differences in Paid Work, Housework, and Leisure* (New York, Greenwood Press).

Sidel, R. (1987) *Women and Children Last: The Plight of Poor Women in Affluent America* (New York, Penguin Books).

Siegel, G. B. (1985) Human resource development for emergency management, *Public Administration Review*, 45, 107–117.

Smith, K. J. and Belgrave, L. L. (1995) The reconstruction of everyday life: Experiencing Hurricane Andrew, *Journal of Contemporary Ethnography*, 24(3), 245–269.

Smith, S. K. (1994, May) Demography of disaster: Population estimates after Hurricane Andrew, Population Association of America (Miami, FL).

Smith, S. K. and McCarty, C. (1994) The demographic impact of Hurricane Andrew in Dade County, *Economic Leaflets*, 53(8) (Gainesville, University of Florida, Bureau of Economic and Business Research).

—— (1995, July) Demographic consequences of natural disaster: A case study of hurricane Andrew, Natural Hazards Research and Applications Workshop (Boulder, CO).

Sorenson, A., Taueber, K. E. and Hollingsworth, L., Jr. (1975, April) Indexes of racial residential segregation for 109 cities in the United States, 1940–1970, *Sociological Focus*, 8, 125–142.

Squires, G. D. (1991) Deindustrialization, economic democracy, and equal opportunity: The changing context of race relations in urban America, *Comparative Urban and Community Research*, 3, 188–215.

Squires, G. D., Dewolfe, R. and Dewolfe, A. S. (1979) Urban decline or disinvestment: Uneven development, redlining and the role of the insurance industry, *Social Problems*, 27(1), 79–95.

Squires, G. D. and Velez, W. (1987) Insurance redlining and the transformation of an urban metropolis, *Urban Affairs Quarterly*, 23(1), 63–83.

Squires, G. D., Velez, W. and Taeuber, K. E. (1991) Insurance redlining, agency location, and the process of urban disinvestment, *Urban Affairs Quarterly*, 26(4), 567–588.

Stack, J. F. and Warren, C. L. (1992) The reform tradition and ethnic politics: Metropolitan Miami confronts the 1990s, in: G. J. Grenier and A. Stepick, III (eds) *Miami Now! Immigration, Ethnicity, and Social Change*, pp. 160–183 (Gainesville, University Press of Florida).

State Coordinating Office for Recovery (1995) *Hurricane Andrew Almanac* (Miami, State of Florida).

Stepick, A., III (1992) Haitians in Miami, in: G. J. Grenier and A. Stepick, III (eds) *Miami Now! Immigration, Ethnicity and Social Change*, pp. 57–82 (Gainesville, University Press of Florida).

Stinchcombe, A. L. (1965) Social structure and organizations, in: J. March (ed.) *The Handbook of Organizations*, pp. 142–193 (New York, Rand McNally).

Streeter, C. (1991) Disasters and development: Disaster preparedness and mitigation as an essential component of development planning, *Social Development Issues*, 13(3), 100–110.

Strouse, C. (1992, 7 December) Public housing Catch-22 leaves Dade tenants in battered homes, *Miami Herald*, 1A, 13A.

—— (1995, 22 January) In hurricane's shadow, a hero's suicide, *Miami Herald*, 1B, 4B.

Sun-Sentinel (1992) *Andrew! Savagery from the Sea, August 24, 1992* (Orlando, Tribune Publishing).

Swarns, R. L. (1992, 28 November) Red tape ties up millions in relief, *Miami Herald*, 1B, 2B.

—— (1993, 3 January) Family bonds battered, *Miami Herald*, 1A, 14A.

Tanfani, J. (1993, 14 June) Off the fast track, *Miami Herald*, 5B, 8B.

Tasker, F. (1993, 22 August) Emotional toll far exceeds physical ruin, *Miami Herald*, 1D, 8D.

Taylor, J. (n.d.) *Villages of South Dade* (St. Petersburg, FL, Byron Kennedy and Company).

Taylor, V. A. (1978) Future directions for study, in: E. L. Quarantelli (ed.) *Disasters: Theory and Research*, pp. 252–280 (Beverly Hills, CA, Sage).

t'Hart, P. (1993) Symbols, rituals, and power, *Journal of Contingencies and Crisis Management*, 1, 36–50.

Tierney, K. J. (1989) Improving theory and research in hazard mitigation: Political economy and organizational perspectives, *International Journal of Mass Emergencies and Disasters*, 7(3), 367–396.

Trainer, P., and Bolin, R. C. (1976) Persistent effects of disasters on daily activities: A cross-cultural comparison, *Mass Emergencies*, 1, 279–290.

Turner, R. H. and Killian, L. M. (1972) *Collective Behavior* (Englewood Cliffs, NJ, Prentice Hall).

Turner, R. H., Nigg, J. M., Paz, D. and Young, B. S. (1981) *Community Response to Earthquake Threats in Southern California* (Los Angeles, University of California Press).

Turner, V. W. (1969) *The Ritual Process: Structure and Anti-Structure* (Chicago, Aldine).

USDC (1993) *Hurricane Andrew: South Florida and Louisiana. August 23–26, 1992*, National Disaster Survey Report (Silver Spring, MD, US Department of Commerce, National Oceanic and Atmospheric Administration).

Vernberg, E., LaGreca, A., Silverman, W. and Prinstein, M. (1996) Prediction of post-traumatic stress symptoms in children after Hurricane Andrew, *Journal of Abnormal Psychology*.

Viglucci, A. (1990, 22 July) Liberty City rises like the Phoenix, *Miami Herald*, 1B, 7B.

von Glaserfeld, E. (1984) An introduction to radical constructivism, in: P. Watzlawick (ed.) *The Invented Reality*, pp. 13–40 (New York, W. W. Norton).

Wakimoto, R. M. and Black, P. G. (1994, February) Damage survey of Hurricane Andrew and its relationship to the eyewall, *Bulletin of the American Meteorological Society*, 75(2), 189–200.

Wallace, R. (1993, 1 April) Liquor use rose after hurricane, experts find, *Miami Herald*, 3B.

Walters, S. (1995a, 2 March) Adventura's success helps North Dade visualize Destiny, *Miami Herald: Neighbors*.

—— (1995b, 27 April) City of Destiny? Supporters rally for change, *Miami Herald: Neighbors*.

Walters, S. and Swarns, R. L. (1995, 6 March) Incorporation fever, *Miami Herald: Local*.

Ward, C. (1990) *In the Eyes of the Hurricane: Women's Stories of Reconstruction*, Women Survivors of Hugo in McClellanville, SC Documentary video, Clemson University.

Ward, K. (1990) *Women Workers and Global Restructuring* (Ithaca, NY, ILR).

We Will Rebuild (1995) *We Will Rebuild! Final Report* (Miami, We Will Rebuild).

Weisman, M. and Moore, A. (1993, July) The role of elected and appointed officials in disaster response, American Society for Public Administration (San Francisco).

Wenger, D. E. (1978) Community response to disaster: Functional and structural alterations, in: E. L. Quarantelli (ed.) *Disasters: Theory and Research*, pp. 17–47 (London, Sage).

Wiest, R., Mocellin, J. and Motsisi, D. T. (1994) *The Needs of Women in Disasters and Emergencies*, technical report for the United Nations Disaster Management Training Programme (Manitoba, University of Manitoba).

Williams, M. (1995, 4 April) Fed-up neighborhoods in Dade County, Fla. want out, *Atlanta Journal-Constitution*.

Wilson, K. and Portes, A. (1980) Immigrant enclaves: An analysis of the labor market experiences of Cubans in Miami, *American Journal of Sociology*, 86, 295–319.

Wilson, W. J. (1987) *The Truly Disadvantaged* (Chicago, University of Chicago Press).

Wolensky, R. P. and Wolensky, K. C. (1990) Local government's problem with disaster management: A literature review and structural analysis, *Policy Studies Review*, 9, 703–725.

Wright, J. D., Rossi, P. H., Wright, S. R. and Weber-Burdin, E. (1979) *After the Clean-up: Long-Range Effects of Natural Disasters* (Beverly Hills, CA, Sage).

WTVJ-Channel 4 (1992) *Hurricane Andrew: As It Happened*, video.

Yelvington, K. A. and Kerner, D. A. (1993, 10–14 March) Ethnic relations and ethnic conflict in tent city: Understanding Andrew's aftermath, 1993 Annual Meeting of the Society for Applied Anthropology (San Antonio, TX).

Yinger, J. (1991) *Housing Discrimination Study: Incidence of Discrimination and Variations in Discriminatory Behavior* (Washington, US Department of Housing and Urban Development).

Zinn, M. B., and Dill, B. T. (eds) (1994) *Women of Color in U.S. Society* (Philadelphia, PA, Temple).

NAME INDEX

SUBJECT INDEX